P9-DMF-358

Bugged

Bugged

The Insects Who Rule the World and the People Obsessed with Them

David MacNeal

ST. MARTIN'S PRESS

NEW YORK

www.stmartins.com

Design by Meryl Sussman Levavi

Library of Congress Cataloging-in-Publication Data

Names: MacNeal, David, 1985-
Title: Bugged : the insects who rule the world and the people obsessed with them / David MacNeal ; with illustrations by Michael Kennedy.
Description: First edition. | New York : St. Martin's Press, 2017. | Includes bibliographical references and index.
Identifiers: LCCN 2017006880| ISBN 9781250095503 (hardcover) | ISBN 9781250095510 (e-book)
Subjects: LCSH: Insects. | Entomology.
Classification: LCC QL463 .M1988 2017 | DDC 595.7—dc23
LC record available at https://lccn.loc.gov/2017006880

Our books may be purchased in bulk for promotional, educational, or business use. Please contact your local bookseller or the Macmillan Corporate and Premium Sales Department at 1-800-221-7945, extension 5442, or by e-mail at MacmillanSpecialMarkets @macmillan.com.

First Edition: July 2017

10 9 8 7 6 5 4 3 2 1

To my sister Kristen & best friend Tony Bellah

If all mankind were to disappear, the world would regenerate back to the rich state of equilibrium that existed ten thousand years ago. If insects were to vanish, the environment would collapse into chaos.

—E. O. Wilson

All those bugs buzzin' 'round your head.

—The Flaming Lips

Contents

Bugged

Introduction

I never appreciated insects until I ripped the guts out of one.

Okay, more like tweezed. In 2011, I had my first insect pinning lesson with a pink grasshopper, aka plains lubber (*Brachystola magna*), in biologist Nick Gutierrez's lab. My friend had caught the pink beauty for me during one of his Arizona bug excursions, and I was invited to help stuff it before it could be displayed in a shadow box. Back at California State University in Northridge, Nick set up taxidermy cutlery and pins to maneuver its six limbs on a Styrofoam sheet. He then casually instructed me to cut the underside of the pink bug to remove its innards.

And it ruined me. For the better.

After I made the incision, I unpacked the organic briefcase like a magician's hat as dark, fetid rot oozed out. What I saw wasn't some would-be windshield splatter. The grasshopper was a micromiracle, containing within itself this sheer complexity of organs and other bits. Its sleek, segmented body certainly put my bag of skin to shame. Performing this centuries-long tradition of pinning—Victorians loved this stuff—awoke my inner entophile. And with all the curiosity buzzing through my head, I wondered: What is our relationship with bugs?

This question spurred a global trek: to New York, where I fed my blood to a jar of bedbugs; to a Brazilian slum, where I released Zika-combating mosquitoes from a moving van; to a beetle pet store in Tokyo; to the blackened, maggot-ridden grounds of a Texas body farm; and to sweltering nights on an island in the Aegean Sea where some locals attribute their increased longevity to a rare honey. What also motivated me to undertake this journey was how our views of insects have shifted in the twenty-first century. Recent technological advances reveal more about insects than ever, like 3-D scans that make it possible to understand bug flight with the goal of building better microdrones; dust-sized computers used to track bee die-offs; and machines separating bug molecules for antibiotics. Even Westerners show an interest in doing what they consider a stomach-churning practice: eating bugs.

All of this potential begins with this invisible force piecing together the planet's ecology.

People tend to think a higher, godlike being runs the show. They're wrong. The real answer is under your shoe. Or flyswatter. Or—parasitically—skin. You know them as the common house pest. But collectively insects are the microscopic lever-pullers calling the shots, shaping our ecological world and plant life for over 400 million years.

Insects compose 75 percent of our animal kingdom. Put into a dessert analogy, that leaves us to share a quarter of blueberry pie with dogs, kangaroos, sloths, jellyfish, marmots, badgers, cockatoos, and the rest of the world's living things. Compared to insects, we are merely blueberry pie crumbs. Pie crumbs in a world of a stable regenerative mass of 10 quintillion bugs. Put into zeros, it looks like this:

Humans: 7,400,000,000
Insects: 10,000,000,000,000,000,000

Insects beat us at the numbers game. For every one of us there are roughly 1.4 billion insects. In 2013, one Reddit user openly asked, "What if suddenly every insect on the planet made it its mission to kill the humans?" A person with the username Unidan then contributed a humorous account of the "insect Armageddon." The contributor used only two species as an example—since it wouldn't take much. Ants, who alone equal our biomass on Earth, could burrow through our nostrils and suffocate us.

I'm happy to report that insects are unlikely to take us on—at least in this way. Their short, sleepless life span is spent doing a myriad of tasks, including the pollination of 80 percent of food plants and the recycling of dead organic matter and waste. (Imagine the stench otherwise.) These are beneficial, multibillion-dollar services keeping life on this planet humming along. The opposite side of that fated coin is a litany of charges: agricultural destruction, home invasion, forest evisceration, and millennia-long diseases and fatalities totaling in the millions across both humans and livestock.

So if we live in a world run by bugs, shouldn't we know how and why they have the influence they do? And, just as importantly, who among our sadly outnumbered lot has had the courage and the smarts to look to them for the grand answers?

These people, it turns out, can be as unusual as the insects they investigate. It's a subculture composed of compelling human beings—an elite few who don't flinch when a bug nears. They see beyond bugs' aforementioned negative attributes—issues, we'll learn, that have been largely influenced by us (human migration, insecticide proliferation). Whereas most of us despise these mini-monsters and get trigger-happy with Raid, the characters I've met are the unique types who, in a way, *converse* with bugs to better define our insect-human ethos and reveal the mysteries of the seemingly mundane. Consider them obsessed.

Frank Krell, a German scientist and walking filing cabinet of information, is one of the key entomological mediators between man and insect. Primarily he deals in excrement. Or, rather, the best traders of scat known as dung beetles. When I visited him at the Denver Museum of Nature & Science, he took me two stories below the museum's ground level and opened a double door into a brightly lit collection center. Rows of sleek white cabinets stacked high, each of them 20 inches deep, brimmed with colorful insects like an archive of hard-shelled candy. Krell slid out a drawer from one to reveal a number of black dung beetles taken from my own backyard of Colorado. He looked at his collection with admiration. "I get paid a little bit," Krell said about his vocation. "So I'm a professional entomologist."

What he meant is that entomologists, people whose profession is to squint at these tiny specks to study their nuances, are largely underappreciated and are driven by passion. Krell's research focuses on how bison turds influence the fauna in grassland ecosystems. When I asked how he found so many beetles, he jovially described his trap and bait system. It comprises a paper plate propped up by a wooden stick going through it, off which dangles a teabag over a cup of water. Picture a patio over which hangs an alluring open bar. The beetles gravitate toward the teabag and get stuck.

"And what's in the bag?" I asked.

"Shit," he smiled. "Human shit, because it's especially smelly."

And just like that, I was sold. My first encounter with an entomologist absolutely delivered. Frank Krell was by far the most intriguing scientist I had yet spoken to in my career as a journalist. And my investigation would only get weirder—but also more awe-inspiring. The point was this: entomologists do things you and I wouldn't typically do in order to decode the marvelous inner workings of insects.

The characters of this story are hardworking, highly intelligent professionals making a difference. I, however, am a geek. Until 2011, my only "interaction" with bugs was a 1990s board game I played as a kid called *Splat!* So, semantic gods forgive me, I'll sometimes say "bug" when referring to insects or arachnids or worms or myriapods (which includes millipedes, for those Nine Inch Nails fans out there).

Above all, this book is a hand-carved cabinet of curiosity. These people, their research, and their stories offer a glimpse into the insect world. The number of scholarly papers written about insects can fill a library, the jargon of which can warrant eyelid-propping specula. But I've parsed through plenty, and I hope you will enjoy whatever has struck my peculiar mind. Like bees, I've foraged the pollen and nectar across this odd little journey, returning to the hive that is this book.

One

A Cabinet of Curiosity

The corpse-colored door hides in plain sight among Soho's posh boutiques. I pass by it at first, missing the "107 Spring" address plaque in tarnished brass. Peering at the buzzer to verify the tenants, I spot the name Stevens. Written below in all caps and in Baskerville font, I spot the word *ENTOMOLOGY*.

Through the safety glass, a dark lanky figure appears at the top of a steep staircase. As he comes closer, I can see he's wearing camouflage cargo shorts, an octopus-emblazoned T-shirt, and strappy hiking sandals. This is Lawrence Forcella, or Lorenzo, who has invited me to this sequestered spot in Lower Manhattan. His stylishly bald head, beard, fat silver earrings, and charisma evoke a modern-day genie—an apropos reference given his daily feats. I say this because after he greets me, we go upstairs to the 400-square-foot room where Lorenzo and a handful of artisans breathe life into dead bugs.

"We process thousands of insects a year," he says as we walk past giant shadowboxes filled with "alive-ish" specimens in the former apartment. This shrine to biodiversity has an inherent *ick* factor. Gentle taxidermists—insect morticians who unfurl the insects' wings and reposition feeble antennae as if to gain clearer radio reception—display butterflies, centipedes, and katydids. In one day they get more intimate with exoskeletal body bits than you and I would in a lifetime.

The department is owned by, and catty-corner to, the Evolution Store—a Victorian naturalist's Shangri-la. Want to buy a fly's life cycle suspended in resin? No problem. In need of an African penis gourd? Pick a size. The clientele ranges from magazine photographers and preppy eight-year-olds spending birthday money on a human skull to Japanese businessmen brusquely pointing at bugs and purchasing the entire lot. And if Lorenzo oversees his team well, nature enthusiasts like filmmaker James Cameron will shell out upward of $10,000 for a display of beetles.

The Evolution Store established a stand-alone entomology department thanks to Lorenzo. Six months after beginning at the shop in 1997, he offered to pin insects instead of Evolution continuing to outsource insect displays. Lorenzo and his taxidermy crew operated in-store. Then Damien Hirst began buying thousands of pinned butterflies in 2005 to create kaleidoscope mandalas comprising a smattering of colors. That same year, Hirst ordered around 24,000 for what became compositions of stained-glass window mosaics. This required nearly 16 butterfly morticians working around the clock; instructions, costs, fumigation, and due dates are recorded in a "Bug Log."

Gradually, taxidermy employees relocated across the street to the store owner's apartment—where I'm standing now with Lorenzo. At some point Hirst, possibly their biggest client, began outsourcing butterflies elsewhere for cost efficiency; still, Evolution maintained its own separate entomology department for 10 years. But when Lorenzo sent me an e-mail about our planned insect anatomy lesson, he hesitantly alerted me that the room would shut down soon due to budget cuts. So I booked a flight. I wanted to know *what* exactly a bug* was.

A taxidermist grabs her time card and punches out as Lorenzo preps this evening's specimen for pinning. The floorboards intermittently creak as I tour the dimly lit space. Metal cabinets near the front door contain plastic shoeboxes of unprocessed raw stock, each with a taxonomic label like ORTHOPTERA, PHASMATIDAE, or HOMOPTERA. The subdivisions and subsets for classification go on and on, and I'd rather not bore you with terms that sound like Hogwarts wizard spells.[†] A rolled yoga mat lies in a shower-stall-turned-supply-closet. Ice in the kitchen's refrigerator usually has to come from a liquor store so it doesn't share the freezer occupants' "dead bug taste." And Lorenzo hunches over a workstation in a room fingerprinted by years of eclectic employees: a curled alien fetus in a jar, a sealed Insect Warrior action figure by Funtastic, Langstroth beehive frames with lived-in honeycombs, and a late-nineteenth-century "Quick Death" pesticide poster.

Under a cone of table lamp light, Lorenzo removes a giant water bug from a take-out food pan its been soaking in overnight. Originally dried, packaged, and shipped from a village in Thailand, the brown, ovular thing no larger than a plump kazoo is now limber and ready to be mounted for purchase. Working for almost 20 years at the Evolution Store has imbued Lorenzo with

---

*Rewind back to Elizabethan times, and the word is tied to ghosts, as evidenced in *Hamlet*, "With, ho! such bugs and goblins in my life." It's also been spelled various ways: the fourteenth-century *bugge* and then a similar German spin *bögge*—root for the night specter *bogeyman* (after all, Tim Burton's Oogie Boogie was made of insects). But why the word was married to insects is hearsay, though some speculate the spook-infested sleeplessness caused by bedbugs in the 1620s.

† But really quick: Insect orders, from which other subcategories stem, have Greek roots. Beetles belong to Coleoptera; broken down you have *koleos* and *pteron*, which together mean "sheathed wing." Diptera—flies, mosquitoes, etc.—translates essentially into twice-winged. Because of their scaled wings, butterflies and moths belong to Lepidoptera. Key word for "scales"? *Lepis*.

the acumen of a furniture salesman. He knows what you want to buy before you do. Collectors appreciate bug mechanics in wonderfully geeky ways, but your average Evolution customer goes for aesthetics, says Lorenzo. *Are you an oak man or do you like walnut? Mahogany? What does your home look like?* Someone with a "strong design sense," he says, might go for the India ink lines drawn on the egg-white wings of a rice paper butterfly. Whereas a customer with tattoos and a nose ring might be interested in a blood-sucking giant water bug.

Should the limbs on tonight's specimen stiffen, Lorenzo's pinning toolkit includes a syringe for injecting warm water to loosen said body parts. He's also equipped with a razor blade to slice underbellies for gut removal and a snuff spoon he finds especially useful for hollowing out goo from tarantula butts. Piped into his computer speakers is riot grrrl band L7, an '80s grunge precursor better known for throwing a bloody tampon at a rowdy crowd. "I admire their grit," Lorenzo says casually as he rubs alcohol over the water bug's back, blotching a paper towel with brown excess grease. Otherwise, the bug would "look like someone put cooking oil on it."

For those who haven't had the pleasure of meeting one, a giant water bug resembles a cockroach with flexing biceps. Its forelimbs have a finger-pinching function used to latch onto frogs and other aquatic animals in ponds or streams and occasionally onto human feet, hence their "toe-biter" sobriquet. Lorenzo prepares one now for today's lesson because New Yorkers tend to refer to cockroaches as water bugs, and it might sell during the summer. "Specifically in New York City they call them water bugs," Lorenzo clarifies, somewhat agitated. "I think people don't want to be reminded of the fact that they have gigantic-ass cockroaches living in their apartments." "Water bug" does sound prettier, I guess. And Floridians call roaches palmetto bugs. "A rose by any other name," as they say . . .

My host, like many entomologists, is stereotypically peculiar.

It's a profession held by people as strange and diverse as the bugs they study. Lorenzo stands out because he's equally brazen and charming, and unlike most in the field, he's completely self-educated.

"I'm not doing this for scientific purposes," he tells me. Certified individuals in the field focus on a specific branch of entomology. For example, a medical entomologist might find ways of stopping disease vectors like malaria-carrying mosquitoes. Or an agricultural specialist might find natural pesticides to combat the forest-decimating mountain pine beetle. "My own specialty," Lorenzo says, "is that I don't specialize." His passion goes beyond ecology. The bugs' intrinsic beauty takes precedence.

Lorenzo's fascination began at age four when he found a dead stag beetle as big as his hand on his friend's driveway in the Bronx—"one of those things burned into memory." When he showed his mom later that day, she took out a box that had a rhinoceros beetle his dad collected while stationed at a Virginian military base. "I realized these guys are all around us . . . From then on I wanted every bug on the planet. Any time I saw bugs I just went crazy over them."

His collection ebbed and flowed for years, eventually falling victim to dermestid beetles. "The great irony of insect collecting," he laments, "is that if you don't properly store your insects, your insects will be eaten by insects." It's aggravating.* A proper collection denotes the date and location of where insects were caught, so, borrowing Frank Krell's analogy, it's like finding your diary eaten by moths. (Although destruction of such evidence of the

---

*Known as the father of American entomology, Thomas Say gathered insects on coast-to-coast excursions in the early 1800s, describing 1,575 new species and amassing one "holy grail" of a collection. But after his death, the beloved cabinet fell into the hands of Harvard College librarian T. W. Harris in 1836. When Harris finally got around to cataloging the insects after leaving them in his barn for an *entire year*, he was shocked to find it did "not contain one half of the species which [Say] has described." The other half rested in the intestinal tracks of moth larvae, beetles, and mites. Submit this mishap to the Annals of Bonehead Moves.

past may be welcomed.) After his bugs disintegrated into dusty mounds, he was discouraged for five years while in art school until he learned about a blow-out sale on bugs by New York–based insect dealer the Butterfly Company. Currently, Lorenzo possesses about 500,000 specimens he keeps in a separate apartment from his own in Hastings-on-Hudson.

Combining his skills as an illustrator and years observing live insects in their natural state, Lorenzo's pieces now seem to pop off the table—were it not for the meticulous outline of pins around their exoskeletons. You can't help but admire the symmetry and anatomy.

On the surface, insect bodies share a three–body segment structure, top to bottom: head, thorax, and abdomen. This makes sense, as the word "insect" also means "cut into." Three pairs of legs attach to the thorax. One pair of antennae perform important tasks like feeling, tasting, smelling, and hearing. And a respiratory system consisting of interconnecting tracheal tubes suck in air through body-segment openings called spiracles. I am not going to go deeper here, but if you did, you'd find a universe of intricacy.*

"The first step with mounted specimens is the pin," Lorenzo says, bare-handedly sticking a pin through the giant water bug's thorax shield, aka scutellum. The average spring steel pin used for mounting is 0.45 millimeters in diameter, with a black enamel finish to prevent rusting and a rounded nylon head. Pushing said nylon head usually requires meager force, but when handling a tarantula specimen with urticating hairs—needlelike defense bristles—it gets painful. Lorenzo learned the hard way. While drying out tarantulas in a 150-degree oven† for prep, he jabbed pins into cardboard without protective gloves, not realizing he was

---

*No joke. Eighteenth-century Dutch lawyer (and insect hobbyist) Pieter Lyonet published a 600-page book with 18 engraved plates that details the anatomy . . . of one bug! Using dissection tools circa 1762, he painstakingly illustrated goat moth larva muscles, which totaled 1,647—three times the amount of a human. Microcosm indeed.

† Next on the Food Network.

grinding urticating hair splinters into his flesh. He shakes his head. "My thumb itched for two years," he says. "Two *fucking* years!" He rubs the spot on his thumb. "It looked like there was pepper under my skin from so many broken hairs."

More pins decorate the water bug sitting on a porous Styrofoam sheet. It rests atop a piece of paper soaked with a khaki puddle of bug juice.

"It makes me pissed off when people view this as creepy," he says, encircling the bug with pins like a knife thrower. "I'll tell ya, the most annoying thing is when I say I'm an entomologist, and people are like, 'Ooo, like in *Silence of the Lambs*,' " and he starts nodding. "Yeah," he sarcastically replies, "I skin women." We laugh, and I can't help but break into singing the goth keyboard synth from "Goodbye Horses" played during the infamous crossdress scene.

The conversation transitions to the John Fowles novel and movie *The Collector*, in which the kidnapper also happens to own a bunch of butterflies. "There's like a lot of negative stereotypes of entomologists and maybe even taxidermists," he says.

"Right?" I agree as he places the final pins around an unflinching leg. "I think it probably has to start with Norman—" *Bates*, we say in unison. I tell him about the unhealthy BDSM relationship with an entomologist in *The Duke of Burgundy*. He fires back with *Woman in the Dunes*, which is a psychosexual romance, again, with an entomologist as victim. My thought is that society in general is not keen on those who dabble with dead things. "The Brits view this very differently from the Americans," Lorenzo says. He mentions another movie, based on the A. S. Byatt novel, *Angels and Insects*, which takes place during the Victorian era. A man processes insect collections for England's affluent. It was during this time when key mediators between man and insect not only answered *what* a bug is, but helped expand the study into the smorgasbord of topics and concentrations seen today. These entophiles built the foundation for entomology. It was done with cabinets. Famous UK banker John Lubbock offered his observation of the period in an 1856 article in the *Entomologist's Annual*: "The present has been called the age of insects; this century at least might be called the age of collections of insects, and not of insects only, for we have collections of almost everything, of shells and stuffed birds, of ferns and flowers, of grasses and coins, of autographs and old china, of Assyrian marbles and even of postage stamps."

These were stored in what are called cabinets of curiosity. The Evolution Store emulates this tradition as does Parisian landmark store Deyrolle, established in 1831, which not only informed the trend but had a part in describing new species that bear its eponym today. Privately owned cabinets of the elite gradually expanded into museums. For those who regarded it as the Victorian version of Beanie Babies, Lubbock gets a bit judgmental but raises a valid point: "A collection of insects which is not studied is of as little real use as books which are not read . . . Yet without collections there could be very little Entomology . . . To describe species so that they may be recognized by other observers is an art much more difficult than would *a priori* be expected . . . [and] if this had been always done, many mistakes and much confusion would have been avoided."

And boy were those early days ever riddled with confusion.

Critical observation lies at the core of bug science. Proper descriptions would aid future research, but those initial details were in dire need of editing. As mentioned in the *History of Entomology*, for example, first-century Roman naturalist Pliny the Elder believed that ticks lack an anus. The origin of bugs was also a headscratcher. Early Asian descriptions of fireflies claimed they developed from "decaying grass." According to Franciscan monk Bartholomeus Anglicus, butterflies were "small birds" whose poop hatched forth worms. The oldest (crude) drawings of insects from wood engrav-

ings are found in the 1491 edition of a Latin natural history encyclopedia called *Ortus Sanitatis*. One depiction of a snail looks like
an eight-legged slug with a yarmulka on its back. Later, in 1602, we
have the first book dedicated wholly to insects, *De Animalibus In-
sectis*.* Finally, by the time of its printing, entomology and, more
importantly, the systematics involved had become an established
scientific study. Archaic yet stylish, the wooden engravings throughout the book are intricate enough for insect identification—more so
than eight-legged snails. The details are a tad embellished. For instance, beehives have hallway designs as though they were a cross-
section of a Swedish hotel, and underground ant colonies, clearly a
diagnostic guess, have an M. C. Escher vibe.

As late as the seventeenth century, philosophers Francis Bacon
and René Descartes believed that insects sprang from decay, with
no reproductive behavior involved. But "spontaneous generation"
was disproved by Francesco Redi in 1668 when he held bugs up to
a microscope, discovering they came from female-laid eggs. (These
prototype microscopes were nicknamed "flea-glasses.") That same
century began to produce scientific illustrations of insects aided
by microscope. Next, Marcello Malpighi further drove entomology as a separate field of study by documenting the metamorphosis stages of a *Bombyx* silk moth. Anatomy began taking precedence
in describing these creatures. Jan Swammerdam detailed the molts
of various insects and developed insect classifications in the mid-
1600s that resemble those still used today. John Ray compiled all
the taxonomic ideas being formulated in "systematical unity" in the
1710 work *Historia Insectorum*.

When I ask Lorenzo Forcella whom he'd consider as the father
of entomology, he says, "I imagine the first entomologist was someone in the Amazon jungles." He walks me to a glass case display-
ing a long-horned harlequin beetle—a red-flamed shell that

---

* A similar book published posthumously in 1634 was written by none other than physician
Thomas Muffet—the man from whom the children's poem "Little Miss Muffet" originates.

Amazon tribes replicated on their war shields. Then he shows me beautiful metallic beetles that tribes strung into necklaces. "We're talking tens of thousands of years! If you start digging into this, it's like peeling an onion."

The earliest depiction of an insect traces back to a drawing on a bison bone from 18,000 BCE. The illustration, by our early Cro-Magnon ancestors, shows a rhaphidophorid cave cricket. The first depiction of a human-insect interaction, however, dates closer toward the first agricultural revolution. A faded painting found in Spain's Cave of the Spider from somewhere between 8,000 to 15,000 years ago depicts a "honey hunter" interacting with a bee-hive, surrounded by a swarm. We can only hope this wasn't the explorer's obituary.

Around 3100 BCE, Egypt's founder of the first dynasty, King Menes, designated the oriental hornet as the symbol of Lower Egypt. It was "probably meant to symbolize the spreading of fear before the powerful monarch," according to 1973's *History of Entomology*. And ancient Egypt's Khepera—a beetle-faced god that was a symbol of creation and rebirth derived from the dung beetles that emerged from rolling balls of excrement—represents the round sun passing over the land. Egyptian soldiers commonly wore scarab rings, and after they died and were mummified, a carving of a scarab was often wrapped in place over their hearts.

In southeastern Cherokee folklore, a beetle creates the entirety of land on Earth from the muddy depths of the ocean. Meanwhile, the Cochiti tribe of New Mexico tells the origin story of a beetle who, while carrying a bag of stars,* accidentally spills them across the sky to create the Milky Way. Crestfallen, he hides his head in shame, which is why beetles look at the ground to this day.

---

*Dung beetles (which can carry 1,140 times their body weight) base their navigation on the Milky Way. Biologist Marie Dacke of Lund University tested beetles' compass orientation skill in a planetarium, fitting tiny hats over their dorsal eyes as they hauled dung. Their routes with an obstructed, moonless starry sky were chaotic, whereas an exposed Milky Way garnered straight-*ish* lines . . . Balls of shit tend to roll unevenly.

The Bible cast bugs in a more negative light. Whenever insects make a cameo, they tend to be associated with the wrath of God (e.g., locust plagues).

In olden days, Athenian women and men wore cicada-shaped hair clasps. Greek children also played with cicadas. Treated as fondly as toys, they were sometimes placed in the children's graves. To help avert the evil eye, an iron locust was placed atop the Acropolis. In nineteenth-century New England, lore has it that dragonflies were deemed the "Devil's darning needle." Dragonflies were said to pay a visit to kids who cursed, lied, or whined while they slept. Children were warned that they would have their lips sewn shut for committing such transgressions.

In terms of systematics, Aristotle is largely considered the first to view entomology as a distinct science by separating what he called "bloodless animals" from the rest. But the study was largely ignored until the Renaissance and the Scientific Revolution, when enough "fathers" of entomology exist to warrant a paternity test. Each field of concentration seems to have these "fathers." Some have even been called the "Mozart or Schubert" of entomology, or, due to their passionate regard, the "insects' Homer." One crucial but underappreciated figure was artist Maria Sibylla Merian, later called the study's "mother."* Of their ilk, the most intriguing was Pierre André Latreille—whose life was saved by a corpse-eating beetle.

An eighteenth-century zoologist and beetle beneficiary, Latreille was born in 1762 in Brive, France, and essentially became the head of European entomology when he took a post at France's National Museum of Natural History in 1827. As a young man,

---

*Maria Sibylla Merian—whose face once appeared on the 500 deutschmark note—was a seventeenth-century artist whose exquisite illustrations of insect life were used to classify new species. Her works were republished into 19 editions by 1771; her illustrations, writes science historian Londa Schiebinger, are "a standard fixture in drawing rooms and natural history libraries." While in New York, I spent time with her famous *Metamorphosis Insectorum Surinamensium* at the American Museum of Natural History. The book creaked open like an old Spanish galleon, and I was seriously tempted to cat-burglar out of the museum with it.

Latreille had quickly gained patrons due to his affability and interest in natural history. Fueled by upper-class support, he had attended college in Paris, scrounging for insects in the streets, armed with the primary tools used by today's capturers—nets, killing jars, forceps, and spreading boards.* By 1792, Latreille, regarded as the "prince of entomology," had befriended notable naturalist Jean-Baptiste Lamarck, who was the first to classify arachnids as separate from insects. Lamarck ushered Latreille into the National Museum of Natural History, where he began to catalog exhibit insects. His work served as the impetus behind his most famous contribution to entomology, *Précis des caractéres génériques des insectes disposés dans un ordre naturel* (roughly translated as *Dissertation of the General Characteristics of Disposed Insects in the Natural Order*). Published in 1796, the work prototyped the current demarcation of insect orders into families. He built off past research: a Dane wrote about insect eyes; a Dutchman, insect antennae; a Swiss gentleman, genitalia; and Carl Linnaeus, a Swede, propelled zoological nomenclature in 1735. Naturalists gradually pieced together what composed an insect, and Latreille incorporated all these ideas while classifying, deriving what other scientists consider real relationships and "families" for them.

Classification is the root of this science. And it almost died in a dungeon cell.

Latreille, who was a priest, seemingly forgot to swear allegiance to the state when the Roman Catholic Church was absorbed into the French government during the revolution. Imprisoned for over a year in Bordeaux, Latreille was sentenced to be executed by drowning. It was in his jail cell that he saw a creeping *Necrobia ruficollis*, awaiting his death like a vulture. The bacon-colored beetle was carnivorous, and frequented decomposing cadavers.

---

* If this sounds like real-life *Pokémon*, it is. Satoshi Tajiri, the game's creator, modeled the addictive Game Boy RPG from his childhood insect hobby. Even nineteenth-century entomology superstar William Spence wrote: "In the minds of most men . . . an entomologist is synonymous with everything futile and childish."

Days later, a physician found the entomology prince crawling frantically on the prison floor "preoccupied" with this beetle, writes zoologist David Damkaer. The physician brought the newly discovered specimen to a friend—15-year-old Bory de Saint-Vincent, a budding naturalist. He was familiar with Latreille and his important contributions to the field, and he knew that this specific beetle was unknown. So the imprisoned entomologist sent a messenger: "You tell [Bory] that I am the Abbé Latreille, and that I am going to die at Guyana, before having published my *Examen des Genres de Fabricius*." Bory's father and uncle managed to utilize what political ties they had, and Latreille was bailed as a "convalescent" under the condition he be returned when asked by French authorities. His cellmates were executed soon after. Today, in the Pére Lachaise Cemetery in Paris, on the base of Latreille's tomb obelisk is a carving of *Necrobia ruficollis* with the inscription: "Latreille's savior."

Thanks to that little guy, our taxonomical lens sharpened over the next century.

The science's true epoch began in 1826. Using Latreille's classification system, English entomologists Reverend William Kirby and William Spence completed the four-volume encyclopedic benchmark *An Introduction to Entomology*. Their descriptions of bug mechanics in both physiology and anatomy have withstood the test of time, finding uses in forensics, pest control, pharmaceuticals, and weapon development. The intricate, factual outlining took over 10 years to produce and must have caused head-splitting migraines. As William Spence lamented: "On [venturing] into *Entomology* we found the most deplorable . . . confusion . . . the same names given to different parts, and different parts called by the same names—important parts without any names etc., etc. so that to make matter for two lines frequently requires anatomical investigations which occupy a day."

The goal of the two Williams' *An Introduction to Entomology* was to make an "attractive portal of economy and natural history" to entice "experimental agriculturists and gardeners" to learn more about the benefit of bugs. In *Bugs and the Victorians*, author and historian J. F. M. Clark further elaborates on their intentions: "Insects . . . provided instruction for the improvement of arts and manufactures: bees and ants were model architects; the insect chrysalis illustrated the beauty and technique of expert lace-makers; and the wasp demonstrated the requisite skills for papermaking."

We'll see later on that William Kirby and William Spence were exactly correct. With a firm foundation established and collectors running rampant, Reverend Kirby went on to become known as the father of British entomology, and by 1833, the Royal Entomological Society was founded. Meetings took place in London's Thatched House Tavern with the society's badge boasting a twisted wing parasite (*Stylops kirbyi*)* named after their honorary life president. Kirby wanted to know, as do I, what secrets insects might hold. "We see and feel the mischief occasioned by such creatures," wrote the wise Reverend, "but are not aware of the good ends answered by them, which probably very much exceed it."

That realization of the good that insects do in the world came later, in the late twentieth and early twenty-first centuries. But it would take a couple of agricultural catastrophes to get there. Thanks to the Great French Wine Blight that destroyed nearly 6.2 million acres of vineyards in the mid-1800s, scientists turned to the entomological findings learned over the past century. In 1854, states began appointing entomologists, starting with New York's Asa Fitch. Congress launched in 1876 what eventually became the Bureau of

---

*The tradition of namesake insects is still practiced. How else would we get the *Scaptia beyonceae*—a golden, bootylicious horse fly honoring Beyoncé? Or for that matter, there's the duly "bandy-legged" Charlie Chaplin fly (*Campsicnemus charliechaplini*), the yoked Arnold Schwarzenegger beetle (*Agra schwarzeneggeri*), and the heavily mustachioed Frank Zappa spider (*Pachygnatha zappa*). For those wanting to create and name a bug of their own, there's the Twitter bot @MothGenerator. Tweet a name, word, etc., and an algorithm spits out a unique, digital moth.

Entomology. As the global population grew, farmers no longer viewed the occasional crop failure as an acceptable loss. Around 1888, Albert Koebele and W. G. Klee introduced the potential advantages of biological control of damaging pests through the parasitic fly *Cryptochaetum* and ladybird beetles, which saved California's citrus industry from scale insects. By 1919, noted one journalist, entomology was no longer "regarded as a harmless and somewhat ridiculous hobby." Fast-forward to 1947, when the United States used captured German V-2 rockets to usher the first animal into space: fruit flies.

The ecological influence of insects is discussed in the modern day in a variety of research studies, news outlets, and, most important, congregations known as entomological societies. About 22 such groups have been meeting for over a century. Mainly these societies are populated by affable folks seeking fellow bug lovers. Still, at times, the groups can be downright vindictive.

Established in 1884, the Entomological Society of Washington (ESW) hosted a vanguard of the field's leaders, like insecticide advocate L. O. Howard, and entertaining members, like German refugee and beetle fanatic Henry Ulke—a man who painted Lincoln's portrait as well as adjourned meetings by playing Wagner piano tunes. The society's purpose? Rubbing elbows and sharing, among other things, cockroach stories. (One involved a nicotine-addicted cockroach Howard cherished.) Among the myths from the early days of the society was the feud between two of its presidents, the lepidopterist John B. Smith and his "rival" Harrison G. Dyar. Although they'd done research together, a "mutual dislike developed" between the two around the 1890s, wrote ESW historian T. J. Spilman, as evidenced by the scientific names of insects they named. In a passive-aggressive move to piss off Smith, Dyar "named an especially fat and ugly moth *smithiformis*." According to Spilman, Smith parried back by dubbing a new genus of moth after Dyar with a pinch of scatologic innuendo. Thenceforth *Dyaria* moths run amok across the world.*

When it came to bumping heads, no ESW member was more outspoken than Alexandre Arsène Girault. Opinions and poetry slipped into his scientific papers. In one from the early 1900s, Girault gives an ESW president a verbal backhand over a disagreement about chalcid wasps:

> *With impunity's gaunt grace;*
> *Ah, come, past coward, lily-livered liar,*
> *Fair-tongued sweet-mouthing unctuous friar*
> *Let's see what's writ across thy face!*

I'm guessing this was the modern-day equivalent to website comment sections turned virulent, otherwise known as flame wars. Lorenzo told me he's seen this online on Entomo-L. This list-serv discussion forum with a design scheme stuck in the 1990s serves as a hub for entomological chatter, and though it's rare, debates do occur between scientists and pest control workers. "If they were in a room," says Lorenzo, "they'd be stabbing each other." But during my time lightly combing Entomo-L, I have only seen the forum function as the entomological societies of yore did when they first began. Members help each other on research papers, sell specimens from collections, and lend out extra Uganda flies from an insect excursion. Among the most popular requests from these hobbyists and professionals is insect identification. Entomo-Lers seem happy to offer their expertise, whereas in some societies—a 1987 survey reported—amateurs get the cold shoulder.

---

* Harrison G. Dyar was a class-A prick. But further digging by Pamela Henson and Marc Epstein debunked the *Dyaria* origins about the "irascible curmudgeon," though the adulterous, unsmiling, "acerbic" grump certainly deserved it. Worthy of note: his secret second family; insulting adversaries alive or dead; "fiery exchanges with colleagues" to the point where they stole his specimen out of spite. The list goes on for this significant yet "tragic figure." But his banker "friend" *did* name a moth genus *Dyaria*.

For your general ID-ing inquiries I suggest using Michigan State University's public diagnostic services. Like a taxonomical WebMD, they'll tell you if that mystery bee in your backyard is sting-y.

During that burgeoning time of entomology in the nineteenth century, taxonomists decided that visual observation alone didn't suffice. Identifying external characteristics and coloration on, say, butterfly wings could easily mislead—the same vexing goofs the two Williams made while writing *An Introduction to Entomology*. As USDA entomologist F. Christian Thompson puts it, "Scientific names are hypotheses, not proven facts." To that end, Lorenzo— sitting in his chair, placing the forty-fourth scattered pin around a bullet ant that could play a minor role in *Hellraiser*—points to a case filled with 30 near-identical butterflies. To the naked eye they are indiscernible. The school of copycats encased inside share black and orange features. But in their pre-metamorphic state, each butterfly came from very different-looking caterpillars. Belonging to the genus *Heliconius*, they benefit from a genetic survival trait known as Müllerian mimicry—the puzzling work of supergenes controlling color-pattern variations that enable poisonous butterflies to mimic each other. That is how these guys raise their odds against predators.

A close examination of butterfly genitalia will help discern what a species is. But the five different nomenclature codes taxonomists use for animal names can be overwhelming. "If you go back and look at publications over the last 20 years," Lorenzo says, "you'd see multiple layers of classification. And they keep moving stuff around because they're not 100 percent sure how everything relates."

DNA classification has already begun to change this. In 2003, Canadian biologist Paul Hebert created a universal data system called the Barcode of Life. Taking genetic material from all of Earth's organisms, researchers participating in the project have now barcoded over 500,000 species. Doing a recent search on newly coded species, I read about the molecular analyses of scuttle flies done by Swedish biologists. Their findings showed that two species described as synonymous in 1920 were in fact entirely separate. To you and me, that might be peanuts. But in terms of future entomological research, the implications of DNA barcoding are significant. Even Lorenzo, who doesn't dedicate much time to taxonomy, is excited.

"From the dawn of entomology to just recently people have been like, 'Ah, I can see the elytra are slightly different on this one,'" he says in a snooty accent. "'We should call this a new species.'" (*Elytra* are the hardened forewings you'd see on a ladybug, for example.) But DNA barcodes will turn past taxonomic characterization on its head, or at least make it more neat.

The Natural History Museum in London is undergoing a similar adaption to the digital age. Scientists recently started to catalog their 80 million specimens from the past 250 years with unique QR codes, with a publicly accessible database that'll progress the understanding of changes in species for future centuries. This system would make the two Williams and that litany of "fathers" writhe with envy.

Back at Lorenzo's Soho Entomology Department, I'm invited to visit him in Hastings-on-Hudson—an hour-and-a-half trip north from where I'm staying with friends in Brooklyn—for a Father's Day bug excursion he's leading through the forest. Of course I accept. It may also be an opportunity to meet some budding entomologists.

Bizarre and complex, insects possess a quality that can make us act equally so. Take for instance eighteenth-century "ardent collector" Pierre DeJean, writes Carl Lindroth in *History of Entomology*. Serving in Napoleon's army, he's rumored to have paused an attack before a battle, dismounted his horse, and collected a *Cebrio* beetle he had spotted. After hiding it inside his helmet, he went on to fight. Though his helmet suffered some damage, he was thrilled to find his "precious *Cebrio* [was] intact." Oh, and they won the battle too.

Then there's Charles Darwin. A big fan of Coleoptera—a student of his drew him riding a large beetle like a kid on a pony—he often toasted wine with colleagues saying, *"Floreat Entomologia!,"* which in Latin basically means "May entomology flourish!" One time, in a scramble to catch scurrying beetles from a tree, Darwin, his hands full, used his mouth to cage a live beetle. He only spit it out once it "ejected some intensely acrid fluid."

Dame Miriam Rothschild's fascination with bugs stuck with her until her death at age 96 in 2005. The naturalist of famed Rothschild banker wealth had an infectious ardor for small things, especially fleas. (Sensible enough. Her entomologist father pinpointed *Xenopsylla cheopis* fleas as the carrier of the bubonic plague.) Entirely self-taught, Miriam became an authority on the great wonders evidenced by an insect 1.5 millimeters in size. By setting up a camera running at 3,500 frames per second, she and a colleague discovered that the powerful jump of those fleas on your cat can hit 400 g-forces—"twenty times the acceleration of a moon rocket reentering the Earth's atmosphere."

David Rockefeller, another wealthy entophile, owned 90,000 beetles. He became obsessed with them at the age of seven. The German model Claudia Schiffer also scrounged in the dirt for bugs as a child, particularly arachnids. Her affinity for terrestrial invertebrates can be seen in the paintings hung on her walls, as well as the spiderweb inspiration behind her clothing line of knitwear. What I'm getting at is there's a juncture where as adults our childlike curiosity, and perhaps intrigue, with creepy crawlies turns testy. It didn't for these guys. Surely, if nature documentaries and YouTube views are good indications, there's a healthy bulk of us still in awe.

The most unlikely hobbyist I've encountered is a heavily tattooed mechanic who worked graveyard shifts at an Oscar Mayer processing plant for 27 of his now 45 years there. By working nights, Wisconsinite Dan Capps was able to dedicate his days to accumulating enough insects to fill over 3,000 handmade cases measuring a yard each. "I have to say, the collecting became almost an addiction," the sexagenarian tells me on the phone in a thick 'Scansin accent. "I may die from the exposure to paradichlorobenzene alone." His claim about the fumigant may hold merit.* Encased in his shadowboxes is a storybook of his life. Capps wrote thousands of letters to insect dealers and other hobbyists listed in *The Naturalist's Directory*, the book used long before online insect trading. His correspondence reached as far as Germany, Japan, and Australia. One exchange with the Natural History Museum in London got him a swallowtail butterfly from 1874. Another exchange in the 1970s, this time with a Parisian auctioneer, scored him what would equate to $20,000 today for a *Mecynorhina oberthuri* beetle "found on only one side of a mountain in Tanzania."

Guys like Capps are rare. His tone over the phone conveys the mellowness of an NPR host. But a photo of Capps from 1969 might as well have been from the set of *Easy Rider*. "At one point, I was a dead ringer for Charles Manson," he jokes, which doesn't help the anti-serial-killer analogies Lorenzo and I made earlier. His biker build was strong and he grew a "nasty-looking" beard. Inked bugs on his arm and shoulder are a fleshy insectarium of memories. They are visible only when he wears the muscle shirts normally donned by motorcyclists. Capps explains that each insect tattoo he's collected has some particular significance: Ulysses swallowtail, Hercules beetle, Crowned Hairstreak butterfly, Eastern tiger swallowtail, *Agrius* moth, and Australian dragonfly are only some examples. Hobbyist friends from various countries continue to visit his house, this walk-in cabinet of curiosity with displays lined floor to ceiling. "This is a monument to my ex-wife's tolerance," he

---

* Toxic vapors of the compound are the same ones you huff in a wooden trunk of mothballed clothes. The fumigant keeping dermestid beetles and other collection nibblers at bay was proven to be carcinogenic by a group of Japanese researchers in 2005. Lab rats inhaling it for two years showed an increase of liver tumors. For what it's worth, paradichlorobenzene is also the main ingredient in scented urinal cakes.

says, reluctantly adding that his obsession may have contributed to their divorce. "I was always careful to acknowledge her and the sacrifices she made letting me spend time on it."

It's partly why he thinks hiding his collection in the basement would be "damn selfish." So he's spent the past 30 years loading cases onto a trailer and driving them to midwestern malls, schools, and education centers from Los Angeles to Florida's Epcot Center. He's always searching for new opportunities to display them even though it's not lucrative enough to allow him to quit his job fixing bologna-slicing machines. His buggy road trips stay on the back-burner for now. (Joyrides in his Wienermobile are prohibited.) Capps hopes that his son Jeff will inherit the collection, his legacy, so it might continue to dazzle those who see it. "The hobby has brought a lot of joy into my life," he says.

Chief among his accomplishments is holding the current Guinness World Record for spitting dead crickets—approximately 32½ feet. The secret, despite what you think, is not saliva. A curled tongue will create a "spiral effect," he told a reporter, and then you aim the cricket's head like a bullet. While touting the ecological benefit of insects, the St. Louis Science Center asked Capps to spit crickets to promote the opening of an IMAX theater. To win tickets, kids had to go head-to-head with Capps. This, he said, was probably the first time mothers encouraged their children to put crickets in their mouths. Anything, he says, to inspire and shape some attitudes. "Many of the insects don't live very long," he concludes. "But their beauty can be preserved for countless years, and it's worth my inconvenience. And I think it enriches my life, and I think it enriches theirs too."

The no. 22 tunnel exit from Grand Central Station opens to New York's cityscape. The northbound train chugs along till the view widens to the lush New Jersey Palisades along the Hudson River like a jungle-draped Great Wall of China. Periodically, you pass abandoned factories, the train's rhythmic sway going *ka-link, ka-link* like the metallic beat of a robotic heart.

I land on the railway platform in Hastings-on-Hudson. With its sun-plumped wooden houses and shops, cat's cradle of power lines, and a diaphone horn to alert the volunteer fire department, Hastings is a primed competitor for a Village of the Year award. A hike up a steep hill takes me to Lorenzo's boxcar apartment where he's eating a bowl of hot applesauce. His place is a low-lit, uncurated museum, with Luna moths, alien photos, newspaper clippings, caged beetles stripping meat off a deer hoof (for the Evolution Store). I especially like the Post-it note cautioning BUGS IN OVEN.

Before heading out on today's bug excursion, we stop at his second apartment. Its Cornell drawers and cardboard boxes house 500,000 or so specimens for his insect-dealing side venture God of Insects. Insect nets lean against the doorway like fishing poles. Lorenzo grabs the one with a time-worn sweep bag and we step outside into the 80 percent humidity afternoon. Sweat beads bunch in the balding cul-de-sacs of my scalp, and I release a guttural noise. "Welcome to Vietnam, right?" Lorenzo says, smiling, his wood-handled insect net doubling as a walking staff. At Hillside Elementary School, our rendezvous point, we find a mix of 22 kids and parents. Lorenzo is surprised by the turnout. An older New England gentleman with neatly combed white hair named Stew Eisenberg tells me how he hasn't hunted insects since he was a Boy Scout, recalling tips from *Boy's Life* magazine. The kid members of this excursion are visibly ecstatic.

"We're gonna get buuuugs!" revels a boy in a Phillies hat leading the charge.

A light breeze bends the tree branches above us as we follow Lorenzo down a wood-planked walkway behind the school and into the forest. He thwacks the tall grass and flowers with his heavy-duty bug catcher. I overhear a dad call him a "net ninja."

After a couple sweeps, he rolls the bag till it's cuffed near the bottom where all the bugs accumulate. "If you're going to be looking for insects, it's more than what you're going to be looking for with your eyes," Lorenzo says, explaining how invisible these behind-the-scenes forces are. "Here is an oak tree cricket nymph. There's a stink bug. Baby assassin bugs. A spider. A leafhopper."

A toddler has a handful of saliva-soaked shirt in her mouth as she gazes at the catch.

"Which one's the assassin bug?" asks one dad, igniting a chain of questions for Lorenzo. "The green one?"

"Yeah."

"What does that mean?" a mom asks. "Do they hurt you?"

"Well, assassin bugs feed on other insects," he answers. "You'll see they have a little proboscis. That's what they use to suck either plant juice or animal fluids. They're one of the richest forms of insect life within the plants here."

Another dad asks his son's question: "What makes a stink bug smell?"

"Um, there's toxins in it for protection. When something eats it, they taste really foul." (Darwin could attest to that.) This inquisitive focus permeates the group. I wonder when kids lose that rubbernecked, enraptured joy. Purdue professor Daniel Shepardson sought to find out more about human-insect interactions by investigating how 120 elementary school children understood bugs. Students across multiple grade levels were asked to draw an insect and explain what it was. In terms of adorableness, you have images with human attributes: "the caterpillar forms a cocoon because it needs a home." And then by the fifth grade, the 2002 research paper notes, physical characteristics (three body segments, six legs) were consistently correct. What Shepardson also found was that children began "to emphasize the negative aspects of insects (e.g. biting, stinging, eating flowers)" as early as the first grade. By age nine, that contention was engrained. Though it might vary culture to culture,* it's clear the beneficial attributes of insects weren't stressed early enough.

Out of all the kids trailing Lorenzo, one in particular sticks out. Una is five at most. Her dad, Ken, hands her a plastic magnifying glass as she kneels down to inspect flowers along the trek. Una's level of intrigue can be felt from a distance.

We reach a clearing at Sugar Pond. Lorenzo gathers the crowd around a waterside log with holes made by wood-boring beetles. "This bench is a habitat," he says. "Here's a predatory ground beetle hunting on the face of this log. They could spend their entire life in this log. That's their world. Think of things in a smaller universe." Lorenzo places the beetle in a plastic vial and passes it around. Una folds her hands around it, leaning her face in closely to observe the thing clawing around inside. Next she offers the beetle up to Stew's wife. She takes it from Una, and holds it at arm's length. The juxtaposition of our different attitudes toward bugs simplified to its core. Stew the ex–Boy Scout nudges her. "Give it a kiss."

We're nearing the end of the trip. Unless you were to, as our bug expert says, "Stop. Stare. And look," you'd miss the elusive wheels shaping the plants and overall global ecology. As a demonstration, Lorenzo leads us to a dead tree trunk. Fresh rainwater seeps through the dirt, intoxicating us with a rich forest-y aroma. The group has tapered off to a select few including a couple boys dragging each other by the ankle and climbing dead trees. "Excuse me, gentlemen," Lorenzo says to the boys, "I have to lift this log. You're standing on someone's house." He unmasks the life below. Seconds later kids are shouting out "worm!" "centipede!" "slug!" "pillbug, pillbug!" Una quietly plays with a leopard slug, laughing as it slimes

---

* In Florence—birthplace of Carlo Collodi, creator of Jiminy Cricket—toward the end of Easter, children can be seen carrying wicker baskets of pet crickets during Festa del Grillo. The insect's charm may go back to the ancient days of Pompeii. While pets were often collected on Monte Cantagrilli, that is, "Singing Cricket Mountain," today many kids opt for cricket toys.

across her finger. I collect a red, plush-toy-looking mite with a Poland Spring bottle cap. "Bugs are more interesting than people,"
one of the dads tells me.

"Look! Look! Your foot's near another one," a kid points out.
There's a ruckus of elations, *ooo*s and *aah*s as more bugs are unearthed from the loose soil and decaying wood. This grotesque
high pervades the science. Nineteenth-century banker John Lubbock mentions it in his 1856 letter about collectors. That those
cabinets, that this crumbly patch of grub-wriggling dirt, are
"storehouses of facts."

Before everyone gets on with their day, Lorenzo concludes the
afternoon's excursion. "Insects are kinda suicidal," he says. "They'll
just throw themselves at the world and hope to survive." Maybe
that's my answer to what a bug is: suicidal. It's a design preprogrammed into their DNA; they've evolved for nearly half a billion
years to be ubiquitous micro-machines pulling off colossal feats
like shaping the world's flora. Some believe this in turn helped
Homo sapiens evolve. After all, sociologically, we have much in
common.

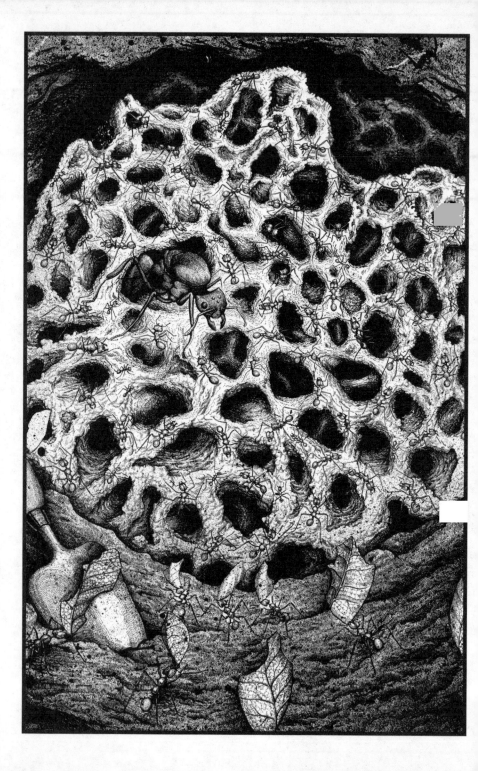

TWO

Buried Cities

Imagine you're an ant. More precisely, a leaf-cutting *Atta texana* ant. Crawling through a bumpy dirt tunnel toward blue celestial sky. You pass by your queen sitting atop a throne of fungus and coworkers carrying leaf remnants over their heads like shark fins.

No, you don't need to smoke whatever's beneath your kitchen sink to experience this. You can navigate a 10-centimeter sample of a real ant colony reproduced in a virtual space on a multifaceted screen. The navigation tool is a Wii remote. Zooming out all the way shows you, the participant, the outer structure of the colony—a view available in the past only through guesswork.

This exploratory, visual map of underground tunnels and chambers is courtesy of a motley crew of computer scientists, artists, and geophysicists led by Carol LaFayette, a professor at Texas A&M. I'm standing beside her collaborator Fred Parke, who helped construct the mini-theater, in a khaki-colored barrack on the university's Riverside campus. Called the *Atta* project, the team used ground-penetrating radar (GPR) scans of an *A. texana* colony in

2006 to create exploratory, visual maps of underground tunnels and chambers. Modeling the GPR data—a sort of "MRI for soil," says LaFayette—with 3-D polygons transformed the otherwise hidden living Texas leafcutter colony into an immersive, algorithm-generated, textured experience.

"What entomologists are interested in are the formation of colonies," says LaFayette. "Their tunnels, their shape, and their distribution can tell them about the behavior of the different species of ant."

This is merely an eight-meter slice of a single colony in Texas. The scope of the leafcutters' abilities—despite being one-millionth our size—are rather gargantuan.* "I've seen excavation sites of this particular species that can hold a three-story house," LaFayette says. An excavated *A. texana* nest in 1960 spanned nearly two football fields. But together, social insects, like ants and termites, can create cities that New York planner Robert Moses would envy, and towers as ornate and functional as the Sagrada Familia Cathedral in Barcelona.

In terms of total population, there are about 10,000 *trillion* of our notorious picnic invaders. *Dorylus wilverthi*, aka African driver ants, exceed the largest of superorganisms known in the world. Each colony holds around 22 million ants (as heavy as a giant Thanksgiving turkey). Over a 250-year course, ants aerate enough topsoil as to add one inch to the world's surface. One colony alone in Brazil—a country where social termites and ants compose two-thirds of the biomass—moved 44 tons of soil. (Comparatively, African termite colonies can aerate over 1,000 tons of subsoil per acre.) One supercolony of *Formica yessensis,* covering 675 acres, with over 1 million queens at work, was recorded in 1979 on the Ishikari coast in Japan's Hokkaido Prefecture. Argen-

---

* That's not to say solitary insects lack design sense. Case in point: as described in a *This Is Colossal* article, the bagworm moth larva construct twisting, pyramid-like "log cabins" to house them during metamorphosis; and the caged orbs of sticks crafted by erebid moths.

tine ant colonies in foreign soil can cordially work in proximity to one another, functioning as a sole "unicolony" that can span an entire state. Hence why some entomologists began aptly referring to these ants in the 1990s as "the Genghis Khan of the ant world." Some more food for thought: rock ants measure the dimensions of a potential new home to confirm its suitability; driver ants create guard walls for food passages; fire ants float atop Brazilian rivers interlocked as a "life raft," reproducing until they find a safe port to recolonize; wood ant colonies can haul up to 100,000 caterpillars a day; honey pot ants sometimes gather a posse of merry (wo)men and pounce on and rob nectar-wealthy ants; and weaver ants use their larvae like construction tools, manipulating them to produce silk to tie* the leaves of their nest together.

This advanced level of cooperative organization in animals is known as eusociality. It is found in ants, termites, wasps, and bees. I'll focus only on the former two here and save the buzzy creatures for later. Such sociability encourages political and ideological parallels to humanity. Touring any human city makes the analogies apparent.

One day in Houston, for example, I saw men in hard hats demolish pavement to replace sewer pipes; police cruisers surveil streets as businesspeople scurry about; diners at a restaurant eat locavore meals on a sunny patio; three men individually approach a woman at a bar; and overlooking it all from high-rise windows, a CEO watches the silent footsteps of workers as they toil below.

Similarly, beneath an anthill in nearby Jessie H. Jones Park, worker ants mine dirt to build windy corridors in an effort to

---

*British naturalist Joseph Bank hitched a ride with Captain Cook to Australia in 1768 and was the first to describe such weaver ant hemming. "Their management was most curious," he wrote in his journal. "They bend down four leaves broader than a man's hand and place them in such direction as they choose, in doing of which a much larger force is necessary than these animals seem capable of." Weaver ants grew more impressive as ant experts honed their observation skills. Like any good contractor, leaf measurements are taken in two-tenths-of-a-second intervals with antennal taps, working the surface like a blind man's white cane.

expand their colony's reach; patrol soldiers surveil tunnels, guarding workers from pests; ants slurp fungus farmed from locally foraged decaying leaves in a dark bulbous chamber; three winged males inseminate a flowery, virgin queen outside the anthill (afterward, she'll burrow into the ground to birth a colony); and overlooking it all . . . Well, the CEO analogy—a role seemingly fit for the queen—is tough.

There is no centralized brain monitoring castes. No cosmic instructions to conquer the world à la Saul Bass's film *Phase IV*. Nor are there the obedient soldier drone ants that Zeus transformed into the Trojan fighters known as Myrmidons. Ants are remarkably, systematically *collective* creatures. And studied closely, the revelations of ant communication have given us faster travel routes to Saturn and may reveal synaptic mysteries in the brain.

From the Gospel of E. O. (Wilson, that is): "Ants are everywhere, but only occasionally noticed."

Across the world there are groups of women and men walking around, heads bent, pondering the fossorial life that teems below their steps. Searching for what E. O. Wilson calls "ecological juggernauts." Without a doubt the most widely recognized myrmecologist (in layman-ese: an ant expert) is E. O. Wilson. Over 16,000 species of ants have been identified. Half studied intently. And never as in-depth as by the famed scientist. What's there to be said about ants? Well, E. O. Wilson and Bert Hölldobler's Pulitzer Prize–winning book *The Ants* carries 7.2 pounds* of information, according to my neighbor's bathroom scale. (The book *Journey to the Ants* is a summarized version of that.) Over the decades, E. O.'s oeuvre has presented remarkable discoveries about animal communication and habitat conservation. The Alabama-born biologist even captured part of his early childhood—a period in time he was nicknamed "Bugs"—in his novel *Anthill*. Through his time at Harvard, E. O. influenced a spectrum of sci-

---

*The approximate weight of 1.1 million average-sized ants.

entific fields, coining such terms as "biodiversity" and propagating previously controversial ideas like "sociobiology." The social qualities of ants and other insects even stumped Charles Darwin. The brood care of honeybees and termites, which among other things involved individual insects remaining sterile for group outcomes, could be "fatal to [Darwin's] whole theory." But then Darwin considered that perhaps natural selection operated on a familial level too. In this case, sterile bugs supporting their fertile relatives made sense.

From the Gospel of E. O.: "The olfactory world of the ants is as alien and complex to us as though these insects were colonists from Mars."

Since 1953, E. O. Wilson and other scientists have discovered more than 20 chemicals ants used to lay trails* or—as French scientist Guy Theraulaz recently found—to saturate the dirt for colony construction blueprints. Myrmecologists have also observed ants utilizing what E. O. Wilson has called 10 to 20 "words" and "phrases" with each other. Colony recognition comes in the form of sensing nonvolatile hydrocarbons in the epicuticle contact pheromones, which look like a "waxy film" coating the ants' body. In an almost tribal manner, initiates are inducted from birth with the distinct colony scent,† a unique fingerprint profile generated internally in the form of these hydrocarbons.

The most innovative ants are the leafcutters *Atta cephalotes,* first described by Danish zoologist Johan Christian Fabricius in

---

* E. O. Wilson demonstrated the potency of the ant pheromone by dipping a birchwood applicator stick into the gut extract and writing his name on a table. People were astonished as the ants filled each letter. Newly discovered chemical codes enabled British artist Ollie Palmer to choreograph his "Ant Ballet" in 2012. Using synthetic pheromones from the University College London Organic Chemistry section and a robotic arm swiveling on a table, he traced patterns for Argentine ants. They readily followed; however, the performance lacked ballet pirouettes.

† To this point, in 1886 Swiss myrmecologist Auguste Forel clipped the antennae off four individual ants—each from different species—and was astonished to see how amicable they were with one another.

1818. Their division of labor is as clear-cut as an assembly line. The process begins in the birthing chamber, where the queen is tasked with laying a modest 30,000 eggs a day. Workers will coddle her for this effort by regurgitating food at her feet. Afterward, the caste system, determined by body size, is physiologically split into soldier ants (with large mandibles utilized the same way Kimbo Slice's monstrous biceps once pulverized contenders); and media worker ants with serrated jaws for fragmenting leaves to later cultivate into fungus. At the end of the cycle is a chamber for refuse disposal, also called a midden, which composts dead ants for fungus fertilizer.

From the Gospel of E. O.: *"The competitive edge that led to the rise of the ants as a world-dominant group is their highly developed, self-sacrificial colonial existence. It would appear that socialism really works under some circumstances. Karl Marx just had the wrong species."*

How such collective behavior evolved is a mystery. Genealogically, ants are related to wasps. They both belong to the insect order called Hymenoptera. The missing link between them was discovered in 1967—and E. O. Wilson nearly destroyed it. When the New Jersey amber piece containing two Ur-ants arrived at Harvard, Wilson eagerly ran to the department and in his overzealous enthusiasm dropped the 90-million-year-old sample on the floor.

Fortunately, the two ants stayed intact. But these primitive ants didn't end up shedding any light on behavior. In 1977, a living species called *Nothomyrmecia macrops* was rediscovered. These ants bore similarities to Mesozoic solitary wasps who began nests in the soil. And more recently, 100-million-year-old fossil records give evidence of cooperative, eusocial behavior in termites and ants. But learning *how* ants evolved into the 16,000-plus species known today requires tracing the subterranean ant nests of the past.

An anthill entrance is visually unassuming. But its true scope reminds me of Hugh Raffles's concluding sentence in his anthropological tome *Insectopedia*: "The minuscule, a narrow gate, opens up an entire world." Unlike the 30-foot-tall thermoregulation termite towers composed of soil, saliva, and poop that resemble the vast Gopuram temples found in India, ant nests are less conspicuous. Yet the underground cities are impressive all the same.

To learn about the evolved sociability of primitive ants, it might help to examine the modern features of these buried colonies. No one has done this better than Florida State University professor Walter Tschinkel. According to Tschinkel, nests come in an array of geometric shapes and sizes, creating "shaped voids in a soil matrix." Using the same plaster* found in an orthodontist's office, Tschinkel fills the nests like you would a Jell-O mold, to later excavate and reassemble the hardened casts. Unearthed, the results vary. A common woodland ants' nest with a workforce of 200, he says, has a "shish-kebob" design, that is, "flattened horizontal chambers" the size of oyster mushrooms connected by a narrow tunnel. But when it comes to the Florida harvester ant, Tschinkel unveils 150 chambers connected by spiraling, helical tunnels boring down 11 feet from the topsoil where the branches cluster and gather like a pixelated jellyfish. It can take 8,000 determined worker ants just four to five days to complete such a structure. Brazilian biologist Luiz Forti created a similarly difficult mold over a decade ago with an abandoned leafcutter colony and 10 *tons* of cement. A month later, the excavation revealed a "labyrinth of chan-

---

* Cast material varies with soil texture: sandy, silty, clay-like. But the ant-enamored man behind AntHillArt.com prefers using molten aluminum on live or abandoned fire ant colonies (an invasive species imported by ship in the 1930s). Dug up, the beautiful yet semicontroversial result is mounted on a wood stand and flipped upside down; assembled, it resembles a chromium coral reef.

nels" extending 500 square feet, designed like a tertiary mycelium network sprouting dome-shaped chambers—a complex achievement that's been compared to the Great Wall of China, but ants smoke that architectural feat. By taking these casts and finding other subsurface fossil nests and scanning them with 3-D morphometrics programs, scientists can show, as Tschinkel writes, "the course of [ants' sociality] evolution through the ages."

Today, we can snake anthills with endoscopes—despite the ants' feisty attacks—and reconstruct makeshift colonies that go beyond Uncle Milton's Giant Ant Farm. In the BBC documentary *Planet Ant*, innovative scientists built a series of dirt-filled terrariums intricately connected with glass tubes like a Rube Goldberg crack pipe in order to watch the cogs and gears of a 1 million-strong leafcutter ant colony.* Setting aside methods used since the 1950s, the *Atta* project was an approach to modeling grandiose colonies that didn't involve plaster, concrete, or aluminum or make the ants homeless.

"But to do it nondestructively, we just don't have the resolution to do much with the data," says Kim Drager, a University of Illinois student. "You can get density versus depth," she says, but the fine details offered by castings eclipse what GPR data can do for now.

Drager's PhD adviser gave her the intellectual freedom to become a hybrid myrmecologist and edaphologist (in layman-ese: a soil expert). Her gung-ho enthusiasm and youth were reflected by the Jackson Pollock–like pants she wore at the Entomological Society of America conference, where she gave a presentation on making casts like Walter Tschinkel and exploring their design implications on the landscape's geomorphology.

---

* John Lubbock's 1876 artificial ant nest is still a charming invention. Comprised of soil between two round glass plates, the widely popular apparatus led to early theories on trail-making and communication. This, of course, coming from the same man who two years later got ants drunk and noted their reaction—for science. "The sober ants seemed somewhat puzzled at finding their intoxicated fellow creatures in such a disgraceful condition, [so they] took them up, and carried them about for a time in a somewhat aimless manner." Lesson: friends don't let friends crawl drunk.

"Soil is a living, breathing thing," Drager tells me. "It interacts biotically, abiotically, it's a carbon sink—without soil, we wouldn't have agriculture. We wouldn't be alive." Still, I'm not too surprised when she tells me little research exists on how insects impact dirt. What's available is summarized nicely in a 1990 paper entitled "The Role of Termites and Ants in Soil Modification." The researchers detailed the increase in carbon, nitrogen, phosphorus, and potassium in areas surrounding a nest due to soil-mixing. The ants regurgitate rich nutrients from below into the topsoil. Ultimately, the researchers found the significance of these soil habitats "difficult to assess." However, the implications, says Drager, could be grand. Arizona State University geography professor Ronald Dorn has measured the calcium-magnesium, rock-forming silicate minerals in several ant colonies for the past 25 years. He found that areas with ant nests increased carbon dioxide capture into rock 335 times faster. That's enough absorption to affect climate change's rise in temperature, which could partially explain how the Earth began cooling 50 million years ago. "It's also species-specific," says Kim Drager. "[Ants] have different modifications that they can do. For example, to keep rain out of their nests."

From the Gospel of E. O.: "The first law of ant colonial existence is that the territory must be protected at any cost."

Their mounds, Drager says, appear to have hydrophobic properties, which prevents the interiors from flooding. While the mound disappears relatively soon after nest abandonment, "the subsurface nest will persist for hundreds of years or more in the soil." And they become conduits for water flow. Modeling how their porosity channels water into the ground may help us understand the rate of slope erosion over landscapes. "Their nests are actually a reflection of their life history," adds Drager. It turns out that they are also possibly a reflection of ours.

One 1990s study of ant behavior actually revolutionized our world with its ant colony optimization (ACO) algorithm. Besides using visual landmarks for navigation, ants leave pheromone trails toward food sources, outlining directions. As more ants take the quicker path, the pheromone (from the Greek meaning "carrier of excitation") reinforces shorter, faster routes, while wanderers find other potential leads and let the useless trails evaporate. An analogous system was programmed into algorithms in the 1990s by Marco Dorigo and two other Italian scientists to solve the traveling salesman problem—a way of graphing the most efficient possible routes to distance sites. "An increase in the computational complexity of each individual," they write, "can help . . . face environmental changes." Their adaptable, "autocatalytic behavior" went on to improve the efficiency of the delivery and transfer methods from billions of different options. This created route efficiency for a bevy of industries, including commercial shipping. It was also used on the 1997 Cassini space probe. Yes, thanks to ants we have faster routes to Saturn.*

Wired with such naturally derived sophistication, these eusocial ants can possibly improve Internet traffic. A study by Stanford ecologist Deborah Gordon and computer scientists found the "evolved feedback-based algorithms" in harvester ant colonies mirrored the Internet's transmission-control protocol (TCP)—a "state-ful" way to manage data congestion online without a central control hub. To elaborate: the TCP algorithm finds "available bandwidth and efficiently utilizes it for the duration of data transfer" to avoid congestion. Should a link break, a "time-out" occurs, which usually appears in your browser window with an exclamation mark.

Similarly: when ants hunt for seeds, the rate with which they return home either decreases or increases the number of foragers leaving the nest so as to maintain steady traffic. "When foragers are prevented from returning for more than about 20 minutes," the study reads, "foragers stop leaving the nest." This system, dubbed

---

*Other ways ACO has proven its worth include (but are not limited to): DNA sequencing, digital image processing, wind farm production, drug discovery, bicycle assembly lines, and the knapsack problem—a resource allocation issue that might expedite the SUV load time before that next family vacation.

"Anternet," has existed for millions of years within ant colonies. According to the Stanford study "Anternet: The Regulation of Harvester Ant Foraging and Internet Congestion Control," their feedback algorithm may be easier to implement than the current TCP. It leaves me wondering what other revelations ants might have in store for us.

One possible avenue may reveal more about the network connection of our own neurons. Currently, Deborah Gordon is working with a computer scientist at the Salk Institute to trace the "network of trails" a certain ant species makes on vines, leaves, and trees in the tropics. As vegetation shifts around, say by a tree branch snapping, the ant network rapidly responds, she says. In this sense it's one network repairing itself within an ever-changing environmental network. There's a similar issue when it comes to synaptic pruning in our minds. "We know [the connections] change in response to changed experience," Gordon tells me on the phone, "so we're looking at the analogy in the resilience of these networks . . . It seems we're going to find a lot of interesting natural algorithms in ants." Engineers might continue the trend of using them in unforeseen ways.

Gordon has spent nearly 30 years watching ant interactions, traveling every summer to Arizona to monitor a growing plot of harvester ant colonies. It's led to some rather controversial theories.

From the Gospel of E. O.: "The exploration of caste and division of labor in ants is in an early stage, and many surprises surely lie ahead."

Since the 1960s, common opinion had it that specialization in body size was directly associated with an ant's job task. "Minor" and "media" workers and "soldiers" were designated to nursing, cleaning, foraging, and guarding, with the occasional reassignment to other duties as they got older. This process is known as age polyethism. But for the past 20 years we've found that, much like us, as Gordon notes, "an individual's behavior changes in response to shifting conditions and colony needs." Organization is more complex than previously thought.

"The daily round of an ant colony is made up of large numbers of brief, simple interactions," writes Gordon in her book *Ant Encounters*. "The outcome is a miracle of fine-tuning." Besides laying down pheromones for food highways, ants communicate via rapid-fire, Morse code–like antennal taps as they stream past each other to "adjust colony activity." It's the same feedback response as with the TCP algorithm. To test her theory, Gordon marked harvester ants in the 1990s, assigning different colors to ones doing certain tasks.* "Then they showed up doing something else," she says. She calls "division of labor"—an idea introduced by Adam Smith in 1776—misleading. "If one unit cannot finish a task," she writes in one paper, "another can take it over with an equal probability of success. This is not possible with division of labor because shoemakers have not learned to make candles." Their complexity goes beyond cogs in a clock, beyond independent parts. The system, like a brain or our daily life, is so much more. Her view on task allocation has yet to be widely embraced by the myrmecology community. Count E. O. Wilson—whom she's met several times—in that category.

"Have you two ever discussed task allocation versus division of labor?" I ask.

She gives a curt "Yes"—and chuckles a bit.

"What was that conversation like?"

Pregnant pause. "Um, well—um. I think Wilson found my work to be a challenge to his work," she laughs. But it's a worthwhile dispute.

From the Gospel of E. O.: "All of ant behavior is mediated by a

---

* Biologists at the University of Lausanne in Switzerland tagged carpenter ant workers with RFID barcode sensors and reaped massive datasets. Tracking their tasks with an overhead camera and computer, the on-screen graphics of their intricate world take the form of an erratic Etch A Sketch. Underneath it all, researchers recorded 9.1 million interactions between individuals over a 41-day period, finding that ants shift careers as they get older, going from nurses to foragers later in life.

half million or so nerve cells packed into an organ no larger than a letter on this page."

And with that comes a question entomologists are solving in a multitude of ways: what contributes to that learned behavior? To get a good handle on all things innate, and the basic mechanics of bug existence and survival, scientists dust off their microscopes and Freudian chaise longues to discover the driving psychological and physical forces of an insect.

One of my favorite questions from the book *Do Bees Sneeze?* is from a third-grade student in Oyster Bay, New York: "Why do flies' eyes look like boom box speakers?"

An apt and cute comparison. Unlike our single vision, some insects have multifaceted lenses. Compound eyes are one of the first bug components that captured our attention. Seventeenth-century scientist Antonie van Leeuwenhoek was the first to view an insect eye's optical array. He wrote in a letter to the Royal Society of London: "What I observed looking into the microscope were the inverted images of the [candle's] burning flame: not one image, but some hundred images. As small as they were, I could see them moving." Two hundred years later, another scientist wrote: "The best of the eyes would give a picture about as good as if executed in rather coarse wool-work and viewed at a distance of a foot." Not much of a view.

With eyes like that, a Van Gogh may not cause the same visual arrest as it does in us. Objects don't register too well for bugs. And though bugs possess trichromatic color vision, they tend to see only shorter wavelengths. But insects' compound eyes, besides giving a panoramic view, allow them to register, say, the speed of a fly swatter much faster than mammals.

Chemically sensitive cells "housed" inside their bodies govern the majority of an insect's taste, smell, and touch. These chemoreceptors may even lie in their legs. The antennae, significant sensory organs for touch and smell, can also be used as airspeed indicators and to detect sound vibrations. Detecting the outside environment? That's due to the mechanoreceptors attached beneath the hairs on their exoskeleton. The hairs also aid in physical orientation.

In addition to Van Gogh, you'll have to scratch Beethoven off the list of bug pleasures. The majority of insects are tone-deaf. Their eardrums, the thickness of which ranges from 1 to 100 microns (the average thickness of human hair), can receive vast frequency bandwidths to, in part, recognize mating calls and assist in echo-locating holla back replies.

Other questions in *Do Bees Sneeze?*—an impossibility due to a lack of sinuses—address plenty of things you may have wondered yourself. Chief brain tickler among my friends, at least drunkenly, was in regard to insects' pain and emotions. How much do they feel when they go *squish*? Is there a certain damnation carved out for malicious children plucking the wings off a fly?

Bugs lack nociceptors you and I possess—nerve cells that alert us when our discomfort is pushed to the brink. It's our emotions that then assign "pain." (Although no conclusive evidence is around, biologists like David Anderson and Ralph Adolphs have shared Darwin's argument that it's possible invertebrates have "central emotion states as well.") The writhing a tortured bug exhibits is a preprogrammed response to survive. As in all creatures, reflex governs most behavior in insects; neurons have receptor dendrites that are connected to the central nervous system, to which impulses travel—a standard locomotive reaction in insects that is known as orthokinesis. A response to air is anemotaxis. To light: phototaxis. (Following the trend of other-axis words, transferring food by mouth [or anus] is known as trophallaxis.) Insects are celestial creatures in that they also respond to Earth's magnetic poles. Evidence of magnetic field sensitivity can be found in termites, flies, ants, and others, improving their foraging direction.

What's exciting is not why insects function the way they do, but how. "The extreme density of neurons packed into a small volume," writes neuroscientist Nicholas Strausfeld in *Encyclopedia of Insects*, "suggests that the largest insect brains have impressive computational power." But describing the parts that make those brains function is difficult. Those microscopic bits of neuroanatomy are hit with the same nomenclature challenges that William Kirby and the early entomologists dealt with.

Within the insect neuroscience community—yes, there is one—brain regions are often described without clear definitions, like "mushroom bodies," "antennal lobes," "lateral triangle," "bulb," and so on.* The problem lies in the difficulty of understanding the synaptic goings-on of the nervous system without a universal lexicon. To solve this, Tokyo students Kei Ito and Kazunori Shinomiya established the Insect Brain Name Working Group. Building a 3-D brain map of *Drosophilia* flies as a framework viewable on x-, y-, and z-axes, they tripled the number of identified neuropils (brain wiring) in bugs. Arthropod neuroscientists have held over 25 meetings to map out the neuron, connective fibers, and neuropil boundaries since 2008.† They've laid out a framework, properly identifying 47 regions, which "comprise the entire brain." Now we have websites like VirtualFlyBrain.org for those wanting to explore an anatomical 3-D reconstruction firsthand.

But what contributes to insects' learned behavior? We've found parasitoid wasps who memorize odors and colors; flies who boost their vision; butterflies who learn to recognize plant shapes. Given their circuit board of neurons, how smart is an insect?

To help answer this, Swiss entomologist Tad Kawecki conditioned fruit flies in the early 2000s. He studied their short-term memory by "infusing" pineapple-like substances with quinine. He left out two petri dishes—one with the tainted pineapple and

---

* We have to admire Henri Viallanes's diligent eye for identifying the optic and olfactory lobes of insect brains. With assistance, he took a lithographic photo—at 400 times magnification—of the bisected brain of a larval fly in 1886. As far as scientific art goes, the bulbous organ is stunning.

† This includes the three-day, Bangalore-held "Maggot Meeting" of 2010.

one with an untainted orange substrate. Following the fruity scents, the flies would land on the quinine-infused pineapple and reject its bitter taste. Later, when they were offered the plates with both fruits on it again, their fly brains were clever enough to associate pineapples with a bitter taste, and gravitated instead to the oranges. Next, Kawecki took eggs from those flies and reared up to 51 generations "favor[ing] associative learning." By the twentieth generation, evolutionary change was obvious. "The experimental flies developed an association between the chemical cue and the medium faster than the control flies," he writes in his paper. They had a "higher learning rate." But with a lack of genetic variation, there was a "slowdown" in evolutionary change; so, regrettably, there will be no dystopian Fly Kingdom mutant overlords for us.

Building on this research, a 2007 study pointed out that flies who "learned faster . . . forgot sooner." Grasping more of that learned behavior could lend insight into how cognitive traits evolve in other species. For example, there's the fruit fly "fight club" at Caltech, where males were bred specifically to become more aggressive. Scientists discovered a substance known as tachykinin in a certain neuron in male flies that is the source of their aggressive behavior. It turns out this chemical is similar to the ones that drive aggression in mammals, like us. Such defensive tendencies are visible across all insect species.

From the Gospel of E. O.: "If ants had nuclear weapons, they would probably end the world in a week."

Fire ant workers will spread-eagle a rival queen's legs while others sting her repeatedly. Captive wasps will gouge predators with their pointy genitalia to break free. Bombardier beetles shoot a jet of boiling chemicals at predators. Some assassin bugs wear the bodies of their kill to blend in among their next victims. Vinegaroon arachnids spray vinegar-scented acid from beneath their whiplike tails. And besides reaching speeds of 5.6 mph—fast enough to temporarily blur their vision—tiger beetles dismember prey with massive jaws and liquefy them into smoothies.

My belief is that, though we cannot naturally turn any species into goop, we share many common traits with insects, or at least find many recurring, analogous ideas. In digging up their microcosm habitats and analyzing their behavior, we can appreciate not only their physiology but their inherent drive, and the engine behind their global contribution.

As we'll see, to know insects is to know a larger function of our life, social or solitary as they may be. But with their aggressive tendencies, hopping into bed with insects can prove to be violent, torturous, and often fatal.

Three

"Even Educated Fleas Do It"

Sexual suicide. Prostitution. Necrophilia. You'll find that the makings of celebrity scandal can also be found in the details of insect reproduction. There's nothing more compellingly absurd, alien, or violent than some bug-on-bug action. Those aggressive tendencies, to put it mildly, were devised as a means of allowing females to spawn safely and reproduce in volume because individually bugs can't build buried cities. But being extraordinary baby-makers assures their critical input in our biosphere—even at the risk of dismemberment.

Some male praying mantises, we've heard, famously risk decapitation after sex via their partner's scythe-like forelegs, but at least their dignity is intact. Others, however, are at times split in half upon mounting female mantises. Brain removed, the nerve cells enable the male's abdomen to continue impregnating her. Male honeybees aren't as fortunate. Drones play a singular role: impregnate the queen. To ensure that, they ejaculate so hard their rocket-propelled gonads burst like a sexual time bomb. Dragonflies and their doppelgänger damselflies—as part of what entomologist

J. E. Lloyd calls a "veritable Swiss Army Knife of gadgetry"—have a hook to yank out sperm deposited by other male rivals. Flea-sized *Neotrogla aurora* females common to Brazilian caves, meanwhile, are endowed with a spiky, inflatable "penis" that docks into males for up to 70 hours of sex. On the prettier side of things, there's the majestic, synchronous, filigreed dance of *Pteroptyx* fireflies, who charm potential mates through photic signaling. Still, males need to possess a *sizable* nuptial gift—a protein-packed sperm sac—to court the ladies. Sometimes female insects simply can't be wooed. Bedbugs get it the worst as evidenced by the number of scars on the female's abdomen, signifying the number of times they've been (forcibly) mated with/gored by a male's bullhorn-tipped penis; a horrifying "traumatic insemination" that makes prison shankings look like amateur hour.

Amber Partridge skillfully prevents such gruesome behavior. She is an invertebrate biologist at the Butterfly Pavilion in Westminster—the nation's first free-standing insectarium—a few miles north of my Denver apartment. As the zoo's head entomologist, Amber performs a variety of buggy upkeep. Chief among them is her knack for refereeing tarantula intercourse. It's important to state that she leads this breeding program not out of voyeurism but for growth study. As far as arachnology goes, these programs are few and far between. It's one reason why Amber has dedicated the past five years to researching tarantulas' life spans, molting, and survival rates. The more she breeds here (five pairs once produced 1,981 spiderlings), the less are removed from natural habitats. When I asked why she does it, I expected some hobbyist response. Instead, it was worrisome: within the next 10 years, she said, the majority of known tarantula species will be extinct.

True or exaggerated, the threat of extinction raises an interesting idea: bugs—those creepy crawlies keeping the world's ecology humming along—might be abundant. But despite their fecundity and ubiquity, or perhaps thanks to it, their extinctions surpass those of any vertebrate. In 2005, biologist Robert Dunn peered over what meager calculated extinction figures existed and inferred in his *Conservation Biology* paper that over the past 600 years there have been 44,000 species lost. Only 70 of them were actually recorded. "The biodiversity crisis," he writes, "is undeniably an insect biodiversity crisis." So, how will they fare in the future?

If Dunn's crystal ball is any indication, it's not looking good. A conservative ballpark predicts that there will be "57,000 insect extinctions per million species on Earth" by 2050, of which less than 1,000 are currently listed as endangered worldwide. Other biologists claim a quarter of all insect species are threatened. Due to bottlenecking from population growth and chemical extermination, those bugs in decline will most likely slip away unnoticed— or, as in one case, while tucked away in a storage cabinet. Such was the discovery made by an entomologist in the 1960s who had neglected to describe a katydid specimen he collected from the Antioch Dunes in Northern California until years later. When he and others returned to the sand dunes looking for a living specimen, there were none. Hence its name *Neduba extincta*. Dunn asserts the causes for such documented extinctions are due mainly to the same ones kicking other animals off the planet—namely, "habitat loss and overharvest." But in terms of ecological function, unlike vertebrates, the impact is dire.

While an endangered mammal is a tearjerker, an endangered insect might be cataclysmic. In 2008, University of Montpellier's Nicola Gallai took a bioeconomic look at 100 crops consumed by humans listed by the Food and Agriculture Organization of the UN and theorized that the worldwide economic pollination value of insects averages $216 billion a year.

Rescue, though seldom, is under way. "Insect conservation," Dunn notes, "remains the awkward 'kid sister.'" Yet nonprofit organizations like the Xerces Society have succeeded in drawing public awareness and, my god, sympathy to endangered invertebrates for decades. Established in 1974, Xerces was named after a radiant blue butterfly last encountered in the 1940s; urban development in the Bay Area likely brought an invasive species that led to its presumed demise. Xerces has spearheaded conservation ever since the insect's discovery, leading coordination efforts to save an array of "at-risk" and otherwise threatened bugs (placed on the Red List of Threatened Species), as well as partnering with various state and federal agencies to monitor and study vulnerable species and help create preserves. As mentioned in the book *Nabokov's Butterflies*, the "first cause célèbre" of John Downey (the lepidopterist who discovered the Xerces blue extinction and later was a Society adviser) was that of the endangered Karner blue butterfly, which was actually first described by Vladimir Nabokov. But it wasn't until 1992, after its habitat declined so much that only 1 percent of the population continued to exist, that the US Fish and Wildlife Service (FWS) listed it as endangered. Normally such a listing might simply impinge on the day-to-day activities of companies, farmers, developers, and landowners. However, the Wisconsin Department of Natural Resources took a different approach to habitat conservation by joining together 40 partners statewide. This, along with Xerces Society's seed-distribution endeavors to create sources for beneficial plants like Project Milkweed—milkweed being the host plant to monarch butterflies—may hopefully pave the way for a 1,500-mile butterfly corridor along US Interstate 35.

In 2015, the Obama administration worked with FWS to protect the iconic* monarch by planting milkweed across the highway stretch. The route also outlines the black and orange butterfly's migratory path from Texas to Minnesota. I should also mention that if bug extinction were the *Titanic*, butterflies and bees would

---

*Before overseeing the assassination of Osama bin Laden, former CIA director Leon Panetta oversaw the sponsorship of designating the monarch butterfly as the national insect. Or tried to, anyway. The bill introduced to the House in 1989 touted the monarch as a native and "unique representative," but only received half as many congressional cosponsors as the "Democratic," "hard-work[ing]" honeybee, even though, as mentioned in the *Lodi News-Sentinel*, it was of "Italian import." The position remains vacant, so write your congressperson now.

be the lofty upper-class passengers claiming lifeboats. Thirty-five percent of the insect species listed on the FWS website are butterflies. Like bees, the reason for this is typically the butterfly's utilitarian values such as pollination prowess and aesthetic pleasure during countryside frolics. But the monarch's decline that began in the 1990s was unavoidable, as butterflies are generally good bioindicators of other problems. Twenty years ago it would've been possible to see 1 billion monarch butterflies migrate to Mexico and cluster, like gooseneck barnacles, on a forest of oyamel fir trees. Today's latest count has them at 56.5 million. If the highway plan works, expect their numbers to quadruple by 2020.

Many conservation efforts also find citizen science useful. To help take census of the decline of fireflies, the Vanishing Firefly Project launched a mobile app in 2013 to track firefly sightings in local habitats, getting an estimated count of fireflies. So far the project's data—coming in from as far as Italy and Colombia—shows how low of a dispersion they have, and how imperative their environments are. But when the number of a surviving species seriously dwindles, and outdoor preservation alone won't suffice, conservation efforts require a more stimulating hand.

That's where the insect Dr. Ruths come in.

Ex-situ conservation is a breed-and-release program. The New Zealand Department of Conservation's success in saving *Motuweta isolata* (tusked wetas) from extinction is an excellent example of an *ex-situ* conservation effort. Wetas are a sluggish species—some, like the *Deinacrida heteracantha*, can reach about 2.5 ounces and can eat a whole carrot. Rescue efforts began in 1993* after invasive rats had largely destroyed the population. By 2001, scientists released ex-situ juveniles, as well as adults fitted with harmonic radar transponders for tracking, into the predator-free Double and Red Mercury Islands. The released adults led researchers to other

---

*Coincidentally, Peter Jackson's visual effects company, Weta Digital, began the same year.

adult wetas. Two years later, island-bred tusked wetas appeared. Now the conservation department is raising other weta species to establish on nearby islands. The Aukland Zoo also used this approach. Extra staff fed hundreds of newly hatched wetas that, if lined along Interstate 35 instead of migrating monarchs, would haunt vacationing families all the way to Corpus Christi. Still, looking like a 'roided-out, long-horned grasshopper bodes much better than being a tarantula.

"We work with uncharismatic species," says Amber Partridge, our matchmaker and biologist at Butterfly Pavilion. "People would rather donate to panda bears." Simply put, we enjoy prettier things. Tarantula eradication proves it.

As a spider, unless you're gifted in calligraphy à la *Charlotte's Web*, you're likely to be greeted with animosity. Myths plague the Mexican redknee tarantula—it is thought to kill horses, cows, and children. And so villagers reportedly pour gasoline into redknees' burrows or rev the accelerator if they spot redknees on the highway. And India's peacock tarantula, a sapphire diamond with eight legs, rediscovered after 102 years, was listed as critically endangered due to the pet trade, negligent construction, and deforestation. These issues also caused the rapid decrease of nine similar species in India—a country where, as scientists unsurprisingly discovered in 1998, "there was not a single tarantula expert."

A Google search also reveals a litany of genocidal images* and videos—things like a Seattle man accidentally burning a portion of his house down to kill a spider.

This blatant contempt is partly why I'm about to watch Amber assist tarantulas' sexytime. Although it's not billed as a conservation-release effort, her study permits a sneak peek into ex-situ breeding. It's funny that one of the least-preserved terrestrial

---

*A particular meme online razzing our overreaction shows a soldier rigged with a flame-thrower setting a field aglow—suitable enough for *Starship Troopers*—with the caption "There's only one way to kill a spider."

invertebrates found its way into the Butterfly Pavilion, effectively sneaking onto those first-class lifeboats.

I must admit that spiders scare me more than death itself. I'm quick to beg girlfriends, neighbors, or even strangers to remove them from my vicinity. I am totally Little Miss Muffet. Apparently, earlier in life, so was Amber.

"I slept in my car for a week freshman year of college because there was a wolf spider in my house," Amber tells me. "I was raised to be *so* afraid of spiders." We're standing in her white-linoleum-tiled office, separate from the Westminster grade school berserkers touring the Butterfly Pavilion. Their zealous screams trickle in. Like a PR lady, Amber is equal parts bubbly and steadfast, armed with a brain library of bug wisdom and a silver Marilyn Monroe necklace charm. Clearly she's gotten over her fear, because stacked along her office walls are shelves of what appears to be dirt and condensation labeled in clear plastic containers. NyQuil-cup-sized water dishes and a mix of curly-haired, Texas brown, pinktoe, and Chilean rose hair (aka "Rosie") tarantulas are inside. All of them were bred in captivity.

"How many spiders do you have?" I ask.

"Oh, my gosh!" Amber exclaims. "I probably have, let's see, there's fifty—plus the one hundred 'Rosies.' Hmm . . ." She trails off, finger on chin as she scans the cramped room. "Probably two hundred tarantulas in this corner here . . ."

My mouth goes dry.

Still, having read through Jeffrey Lockwood's *The Infested Mind: Why Humans Fear, Loathe, and Love Insects*, I'm reluctant to call myself arachnophobic. My instant reaction to avoid unpleasant situations models that of the opossum—I yawn, get drowsy, and bow out. Yet there are roughly 19 million entomophobes in the United States who might experience the "automatic and uncontrollable" trembles, shortness of breath, or "irrationally exaggerated responses" as summarized by Psychiatry.org. It's an age-old fear capitalized on since the late 1800s, as shown by magician Henry Roltair and his "Spidora the Spider Girl" sideshow illusion—a spider with the head of a woman.* Entomophobia had also been an "emotionally traumatiz[ing]" inspiration for Salvador Dalí, who in one buggy confrontation with grasshoppers leapt out of a window. (Grasshoppers *sprung* their way onto many of his paintings too, including *The Great Masturbator*.) And that fear was nearly exploited by the CIA as an "Orwellian strategy" during al-Qaeda terrorist interrogations by threatening to place entomophobic prisoners in a dark containment box with a "stinging" insect, writes Lockwood. In reality, a 2002 memo from the US Justice Department proposed that they'd "place a harmless insect in the box, such as a caterpillar."

Luckily for entomophobes, exposure therapy is available. Methods vary from hypnosis to physical confrontation to virtual reality simulations, like the mid-1990s, first-person hit *SpiderWorld* developed by University of Washington professor Hunter Hoffman, which helps desensitize[†] patients. After participants complete the simulation, they view all spiders as gentle. And it's not a trick: for the most part, spiders are. Amber Partridge occasionally lends her Rosies to psychologists, aiding patients by strengthening their mental fortitude with tarantula knowledge, perhaps to the point of getting them to hold one.

Now some tarantula Sex Ed 101: each developmental stage of an arthropod is called an instar. For example, tarantulas undergo molts where they shed the exoskeleton of their former selves, maturing around ages four to seven. Picture an eight-fingered hand working its way from a leather glove. After this ultimate molt, males can now breed, and many are equipped with protruding

---

* Paging Kafka.

† In 2014, an alternative and by no means *extreme* solution to ridding phobias was undertaken in the form of a left temporal mesial lobectomy.

tibial spurs on their front legs intended to hook the female's fangs while the male crawls beneath her for insemination. Otherwise, she'd dig her fangs into his noggin.* The spurs later prevent males from escaping their postultimate molt—a tarantula's final shedding. Unable to jump out of their skin, they die from dehydration. As part of her study, Amber ushers them through that phase, noting how "super awkward" they become. They're confused. Their gait is off. I imagine the human equivalent to this senescence would be expecting to die at 65 and waking up at 90. Longevity in the arthropod kingdom really goes to the females, especially since some hobbyists have said female Mexican redknee tarantulas live to see 50. The males need only to mature enough to contribute their chromosomes. "You get to a certain age and you're like, 'Okay, I really want to go to the bar tonight,'" Amber says, laughing through her nose. And after their jollies are had, it's checkout time.

For today's demonstration, Amber calls upon Greg Kinnear.

Initially, Amber's spiders get numbers, and, though rarely, accidents will happen. When I visit Amber's lab, Rosie no. 115 is recovering in the "Non-Working Rosies" sickbay. Successfully impregnated spiders receive actual names. Same if they encounter—like our male specimen today did—a celebrity. Besides loaning out Rosies to psychologists, Amber is summoned to Hollywood to wrangle insects on film sets,† as she did for *Heaven Is for Real*. She

---

*Affectionate love bite this is not. For that we turn toward Costa Rica's peanut-headed lanternfly, colloquially known as *la machaca*. Legend portends its deadly venomous sting leaves its victims with but one cure: have sex within 24 hours. Kudos are of course owed to the centuries-old bullshit artist who popularized this harmless insect across South and Central America. For a real insect-derived sexual stimulant, I recommend the Brazilian wandering spider. A single bite will give a man an hours-long erection. If it replaces Viagra, I'd like to propose an ad campaign designed by the Monster Energy drink people for a pill called "Boner Venom."

† The go-to wrangler for decades has been entomologist Steven Kutcher, whose wide range of credits appear in *Jurassic Park*, the *Spider-Man* trilogy, *We Bought a Zoo*, *Fear and Loathing in Las Vegas*, and, most famously, *Arachnophobia*. Kutcher employs a

convinced the movie's main star, Kinnear, to hold a Rosie—the one we're now attempting to mate, thus convincing me that heaven is indeed for real.

I ask if we should put on some Al Green. "We always joke about that," she says, half expecting the question. "But no. No Barry Manilow." (I think she means Barry White, but "Copacabana" works too, I guess.) Our resolute matchmaker sets down a cage on a metal table fit with a partition between Kinnear—the tarantula, not the actor—and a female Rosie, which is perforated in order for their pheromones to travel through the terrarium, permeating and enticing each other. Amber must now steer the two using a green paintbrush she keeps in a pencil jar. Looking around at the 300-plus spiders, it's obvious she is good at her job. She's bred 23 pairs, 16 of which successfully had offspring. On average, her spiders mate within 15 minutes. So, she removes the plastic divider. The starting gate opens! And—

Nothing. Neither of them budges.

Kinnear emerges from his trailer—a tipped-over clay pot—his dark pink velvet asterisk of a carapace gleaming. The stage is already dressed with a web laid across the pebbly floor laced with thick ribbons of sperm, some of which Greg Kinnear—of Westminster, not Hollywood—has already sucked up into quill-like appendages near his mouth called pedipalps. But the atmosphere is tense. Amber guides the female toward him. Normally, in the wild, a male approaches a female's burrow and politely taps on the edge. If she doesn't come out to mate with him, he'll try a burrow next door. Some male Rosies are brave enough to trespass inside. "But that could result in death," Amber warns.

Hoping not to end Kinnear's career, Amber delicately guides

---

variety of tricks to guide and place insects, from cooling them to slow their metabolisms to constructing walking paths out of threads. It's Hollywood smoke and mirrors at its smallest.

him, using the brush to rub the pheromones off his pedipalps and onto the female's. She nudges the two closer. They're coy.

"Kinnear doesn't want any of this," she says, disappointed by their apprehension. At one point, the female lifts onto her hind legs, presenting herself for insemination—but Kinnear hides in the corner. "He's camera shy. He's not a very good actor," she says, visibly upset. "She may be a little aggressive and he's nervous." It could also be that she's giving off a gravid pheromone telling him she's already pregnant.

Instead, we opt for the spider that originally frightened Amber as a college freshman: a wolf spider.

We do an about-face in the small room and check on a pair in the terrarium. "Hi, you're very pretty, honey," she pacifies the fuzzy female. "We have a divider in here, but he decides to climb over it every night," Amber explains. Normally, they'd receive separate homes, but caging both for a week to incubate pheromones makes for faster breed times. But something's amiss.

"Humph . . ." Amber is stumped, and uses her brush to probe the male's side of the divider.

Oh, no, I think.

"Did you eat your boyfriend?" she tranquilly asks the female.

His side of the court is empty. There's zero trace of any limb-y morsel.

"That's unfortunate . . . You look like you ate him." Dismayed, Amber hardly speaks as she sets aside the now lone wolf spider and grabs a pair of pinktoe tarantulas. We both brim with hope, noticing the sperm web draped on the dirt floor. But the once-amorous mood here at the Butterfly Pavilion has turned anxious. The pinktoes also get cold feet.

"I guess—" she sighs. "This is ridiculous. This is seriously like herding cats! I've never had this much trouble ever—ever!"

Omens apparent, we decide to pair up different spiders and let the pheromonic anticipation simmer for a week. Spoiler alert: we will succeed, and, yes, I get hands on.

Snippets of the 1964 research paper by professors W. H. Whitcomb and R. Eason entitled "The Mating Behavior of *Peucetia viridans*"—about green lynx spiders easily mistaken for crawling boogers—verges on fervent: "They touched each other rapidly and repetitiously . . . Often, he twirled her by only touching the thread, but at other times, he touched thread, legs, and body . . . He began to drum on the tip of her abdomen with his palpi and the tarsi of his forelegs, his body often trembling concurrently."

Scientist and author of *Sex on Six Legs* Marlene Zuk comments that some of these insect mating papers "sound like what would happen if Danielle Steel were an entomologist." Reproductive behaviors in insects appear so odd and incapable of characterization that scientists who up their wordsmith game deserve a pat on the back.

The *P. viridans* action gets spicier: "The female then bent into a shallow 'U-shaped' position . . . With lightning speed, the male thrust his whole body forward and jabbed at the female's epigynum with his palpi, first one and then the other." Using 16-millimeter motion-picture film to shoot 40 pairs of spiders housed in ice cream pints, the entomologists found that the time length for courtship averaged 11 minutes, and mating lasted 10 minutes.

Former Lancôme model Isabella Rossellini entertained us with buggy sex in half the time. In 2008, when the Sundance Channel was looking to air five-minute segments between programs, the Italian entophile figured the best way to "capture people's attention" was raunchy, educational descriptions of insect sex. The resulting combination of plush and cardboard costumes for the segments called *Green Porno* stemmed from Rossellini's childhood dream to make films about animals. Seemingly sewn together by some beatnik *Sesame Street* prop master, the costumes set the stage for Rossellini to run amuck. She humped praying mantis and fly statues, and dressed as a tubular earthworm imparting anthropomorphic winks to the camera, saying: "To have babies I need to mate with another hermaphrodite—in the *sixtynine* position." Donning compound eyes in complete dragonfly getup, she explained how "first I will clean her vagina to make sure she would only have my babies. Then we would copulate."

Good insect sex material seems endless. For instance, University of Canterbury researchers found *Portia labiata* female jumping spiders, with their misleadingly adorable convex eyes, usually eat their mates after sex, executing a sudden "twisting lunge" with fangs drawn while chomping down on the male's head. Bizarre insect mating facts like this abound. Scorpionflies snack on fresh prey while in bed. A Silverfish male sets up sperm packet traps with thread and lures a female to crawl underneath as it drops the payload into her. Necrophilic flour beetles are quick to copulate with their dead female companions.

Smaller male rove beetles mimic females by having sex with other males to finagle their way to reach the ladies. Such documented male-on-male actions, Marlene Zuk writes, are sometimes mistakes that insects—about 117 known arthropod species—are willing to make rather than miss out on actual reproduction opportunities. Research also shows that male fruit flies that engaged in homosexual practices had better chances of scoring with females. Similar sociosexual behaviors were observed in silkworms back in 1909 by Italian entomologist Antonio Berlese. And a recent survey done by two researchers in Mexico pushes the hypothesis that lepidoptera males exhibit copulatory attempts "to inflict damage to sexual competitors"—habits also seen in bedbug threesomes. The topmost male injects his sperm into a male who is already ejaculating into a female on the off-chance his squigglies will flow into her, supporting the idea that it is not only survival of the fittest, but rather of the most deceitful.

Affluence works too. Dance flies, like many insects, require payment for sex with gifts, such as small prey or nonedible tokens such as pebbles, leaves, or silk balloons filled with food or left empty.

A good deal of moths, like the European corn borer, willingly put out for any species releasing pheromones. But those moths, like many other species, suffer for their promiscuity. Entomologist James Wangberg's book *Six-Legged Sex* alludes to an STD within "5–10 percent of all insects" called Wolbachia. Some studies even estimate the number at 20 percent, tiptoeing closer to human STD numbers. The transmitted bacterium, which lives in the female's eggs, makes procreation possible only with other Wolbachia-carrying males. It's either that or reproducing parthenogenetic carbon copies of themselves, aka virgin birth, resulting in only female offspring.

As absurdly complicated* as bug mating rituals can be, their abundance and variety proves that they've worked for over 400 million years.

A pair of one-and-a-half-inch froghopper insects, "belly to belly" in a standing missionary position, are freeze-framed in coarse sedimentary rock dating back to the Mesozoic Age. The bottom of their abdomens connect, the entomologists write, his aedeagus (essentially penis) entering her bursa copulatrix (basically vagina). That makes the 165-million-year-old fossil the oldest record of insects doing the dirty. Found in northeastern China in Daohugou Village, the species *Anthoscytina perpetua* was named by Beijing researchers from the Latin *perpet,* meaning "eternal love, in reference to this everlasting copulation." The amorous image caught the interest of

---

*Afterthought: perhaps the strangest bug sex goes to those humans with formicophilia. Sufferers of this rare deviant sexual behavior can only experience arousal and orgasm from ants, snails, cockroaches, or "other insects creeping, crawling, [and] nibbling" their skin and genitalia.

those studying the evolution of asymmetrical insect genitalia, since it shows more evidence of shared similarities from that time with modern bugs.

Today's insects have plenty in common with their prehistoric ancestors. In fact, certain stick insects have cloned themselves for over 1 million generations. A team of Canadian scientists traced the lineage of several *Timema* stick insects using DNA analysis to find when they became distinct from their elders, revealing they maintained asexual reproduction since the Mesozoic age without the introduction of new genetic diversity from males. But it's by going back over 300 million years to insects' Paleozoic origins that we can better illustrate how their, as biologist Michael Samways puts it, "evolutionary resourcefulness . . . in prehistorical times" applies to future climate change.

Within the Paleozoic era was the Carboniferous period, aka the "Age of Cockroaches." If you were magically transported there, you'd be surrounded by lush arborescent plants, far-reaching coal swamps, and eight-foot-long arthropleuras—extinct ancestors of modern centipedes and millipedes—quietly scurrying past your feet like serpentine surfboards. Because of the higher temperatures, there were an astonishing 4,500 species of cockroach. But during the Paleozoic era, you'd also have to dodge buzzing dragonflies large enough to carry frogs and mesothelae spiders capable of enveloping your entire face. The air during this period was composed of about 35 percent oxygen, as opposed to today's 21 percent, which physiologically required a bigger respiration system in the biome, hence the gigantism. Over the next couple of periods, temperatures cooled and rose. Mass extinctions occurred, save for 8 of the then 27 insect orders. Those 8 are now relatives of today's bugs. Eighty-four percent of current insect families are the same as 100 million years ago.

Which brings us to the increased concentration of carbon dioxide in our atmosphere today, raising the global average surface temperatures by half a degree Fahrenheit from 1906 to 2005, according to NASA. The trend is not lost on conservation biologist Michael Samways. He and other biologists theorize these ectotherm insects, "sensitive to temperature changes," will migrate to "geographical ranges closer to the poles" and flee to higher elevations, upping the diversity found in those regions. An example of this shift can be seen with the Comma butterfly, which over the course of three centuries has relocated itself 100 miles northward in England. And potato leafhoppers arrive 10 days earlier from the Gulf than compared to summers in the 1950s. A 2009 paper exploring these thermodynamic effects found that such varied temperatures would "constrain the maximal performance of organisms," favoring insects with thermoregulation adaptations, aka endotherms. Meanwhile, a recent census by researchers Timothy Bonebrake and Curtis Deutsch shows insects in topographically diverse tropical regions like South America and eastern Africa "are predicted to be more physiologically capable of tolerating changes in temperature." Figuring out the mechanisms for wider tolerance in bugs at higher elevations, they say, will require further analysis.

Also of note, says Samways, is how associations changed with certain crops. The increase of carbon dioxide will "likely result in greater carbon-to-nitrogen ratios in plant tissues . . . stimulat[ing] greater feeding activity in some insects." Reproduction of the Texas field cricket decreased with higher-than-average temperatures, one study found. Their survival "hinge[d] on food availability." The Texas cricket researchers at Dalhousie University believed that if their temperature experience was applied to other insects, climate change would favor agricultural pests in temperate zones, *increasing* their reproductive rates when there's enough food to sustain their survival.

What effect would an even more dramatic change, like a nuclear holocaust, have on the insect population? Should one occur, I'm sure you imagine a future beyond just Twinkies and cockroaches, right? Timothy Mousseau aims to find out. The University of South Carolina professor is predicting how radiation will affect animal life over the next couple decades. For nearly 20 years, his team has surveyed the contamination in Chernobyl's exclusion zone to trace radionuclides and their effects on humans. In 2011, he began inspecting fauna in the Ukraine and later comparing them with insects and other animals in Fukushima. Let's just say it's a far cry from the nuclear monster movies of the 1950s. Biological life has changed. The pulse is faint but also revealing. Following radionuclides—isotopes that can emit cancer-causing gamma rays—Mousseau has witnessed firsthand how natural populations are affected by this shift in mutation selection balance and how it affects the evolutionary process and, ultimately, adaptation for survival.

What sparked his initial interest was a firebug. He and long-time colleague Anders Møller were attending a conference in the Ukraine during the twenty-fifth anniversary of Chernobyl and exploring the nearby city of Pripyat. Crawling among the abandoned buildings, the rust-coated Ferris wheel, and the other eerie elements of the city evacuated by 50,000 people were mutated firebugs. "They're cryptically colored in this red motif on their backs, sort of resembling an African face mask." Møller grabbed one and said, "Look, Tim! This one's missing an eyespot!" Some dots looked fused together. As the researchers expected, their abnormalities were related to the background radiation. Møller and Mousseau have been taking samples of fauna ever since. Together, they trudged these zones led by the static ticks of Inspector Alert Geiger counters. A couple of sprays of a water mister revealed odd spider webs as imperceptible as the radionuclides surrounding them.

"We noticed many of the webs were peculiar looking," Mousseau says. In the photos he's sent me from Fukushima, the cobwebs wilt asymmetrically, some with a gapping pupil like a lazy eye. "We were new to Japan, and you know, maybe Japanese spiders aren't as good as making symmetrical webs as spiders elsewhere," he laughs. "The bottom line was there are a higher frequency of abnormally structured web architecture—less uniform—in the radioactive areas. And this is true in both Chernobyl and Fukushima." They hypothesize the oxidative stress associated with radiation affects the neurological development in spiders, causing the misshapen webs. That aside, spider populations seem to increase with radiation—possibly a result of weakened, irradiated prey. Meanwhile, other insects decrease. On one particular trip, he hefted four UV-light-emitting moth traps powered by car batteries to collect kilos of moths at night. He found an abundant difference in bounty between clean areas and places shrouded by radioactive fallout.

Like the firebugs that first entranced the two men, the mutations* are not only obvious but also reproducible in laboratory settings. This is crucial when trying to study the heredity of abnormal traits. Malformations from radiation experiments date back to the early 1900s. Claude Villee used X-rays on flies, producing leg-like antennae or an extra palp growing from an eye. After the Fukushima meltdown, researchers in Okinawa re-created the same effects in Japan's most common butterfly, the pale grass blue, by feeding them irradiated leaves. The results were menacing. Rates of "forewing size reduction, growth retardation, high mortality rates . . . color pattern changes" grew over generations, which implied an "accumulation of genetic damage."

At the opposite end of that gene-churning spectrum, the Antarctic's chironomid midge (or sandfly for you Midwesterners) can dehydrate itself to survive at 5 degrees Fahrenheit without ice crystals destroying its cell membranes. Its relatives in the Nepal Himalayas can bear going down to 3 degrees and still happily

---

*Science illustrator Cornelia Hesse-Honegger documents anomalies from copious radiation. Her harrowing sketches of insects near nuclear waste facilities and fallout zones, like Three Mile Island, capture the insects' haunting yet beautiful flaws: dented eyes, lopsided wings, stubby antennae, and missing legs. As she writes in *After Chernobyl*, "We cling to images that do not correspond to changing reality."

graze off blue-green algae growing on glacial ice. Ice worms—inch-long, black annelids—burrow into solid ice like earthworms do into soil. Finding them is difficult. Especially as melting ice pushes them further into Alaska snowfields. "We're interested in how they've adapted to ice," says Dan Shain, who's spent 15 years tracking these worms that NASA loves so much. The space agency funded $214,000 for research to look into how life might be sustained in icy habitats across the galaxy. The mystery, he told one reporter, is figuring out why the worms have an enzyme that "skyrocket[s]" their metabolism—specifically adenosine triphosphate (ATP), which transfers intracellular energy. Due to such intense energy, ice worms "thrive" in below-zero climates and disintegrate at temperatures over 40 degrees. Shain and his colleagues believe ice worms' tolerance to cold is related to a subunit of the ATP synthase complex. This "throttle" protein has been modified by ice worms "in an unusual way," Shain said in an e-mail, since the protein adds an 18-amino-acid extension that increases the worm's energy levels.

Finding such adaptive abilities puzzles and amazes nearly every biologist. I'd be interested to see how hardy ice worms would be in a lab setting. How many generations it would take to adapt to 40-plus-degree habitats. I know after talking to Marlene Zuk that witnessing those changes in person can be unnerving. That's what happened in 1991 when Zuk crawled over a Hawaiian grass field at night, looking for crickets that might contain parasites, headlamp illuminating an abundant amount of *Teleogryllus oceanicus* crickets expressionlessly staring back at her. Only a few of them were singing. By 2003, they were chirpless and bountiful.

"If you know anything about crickets," says Zuk, "then calling is the sort of thing they do." Male crickets have a stridulatory apparatus on their wings with teeth and a scraper, known as flatwings, that produces sound when rubbed like a bow to a fiddle.

The emitted chirps* act as a beacon for potential mates in the area. As she explains in a paper entitled "Silent Night," a scanning electron microscope's micrograph of the flatwings in *T. oceanicus* showed the same apparatus as other crickets, but at a much reduced size. Nature pressed the mute button. These chirpless crickets, she says, are an example of rapid evolution.

Those crickets that Zuk discovered in the field at the conference, *T. oceanicus*, were originally from Fiji, Tahiti, and Samoa but had been on Hawaii since 1877. Some speculate the crickets were introduced by Polynesian colonizers 1,500 years ago who, folklore has it, believed crickets were their deceased relatives. But their move to Hawaii brought them a new experience that has affected their evolution: a parasitic fly hones in on the crickets' broadcasted mating call and deposits mobile larvae on it. Larvae then develop inside and, as Zuk's students say, "eat all the gooey bits." After only 20 generations, the crickets changed their gene frequency to radio silence. It wasn't just behavioral—they were "physically incapable of producing sound." Visit Kauai today, and you'll find about 90 percent of *T. oceanicus* males are hush-hush. On Oahu? Half won't make a peep. What's really perplexing is that while *T. oceanicus* is the same on both Kauai and Oahu, an analysis shows the single gene enacting the chirpless wing morphology mutation is different. So in order to continue reproducing, the crickets utilize a "bait-and-switch strategy" by lingering near males of the same species that can still call, and, in a cold-hearted, frat-boy gesture, swooping in once a female appears.

Occasionally a species that's gone extinct will turn up resurrected decades later. This phenomenon is aptly called the Lazarus effect—an act of perseverance exhibited by the tree lobster last seen off the coast of Australia in the 1920s but that is alive and well today.

---

* Enamored by ancient Chinese traditions, Mr. Fang Liao professionally breeds and installs live cricket ensembles, and has learned many secrets to their musicality, like dripping resin on their stridulating legs and wings for lower octaves. Literally rosining up that bow.

Ball's Pyramid, near Lord Howe Island in the South Pacific Ocean, doesn't feature rounded contours but rather is a pointy remnant shield volcano. In 2001, a team of surveying researchers found moist, green frass (i.e., insect excrement) in a shrubby patch on Ball's. The frass belonged to the *Dryococelus australis* tree lobster of yore. Sure enough, later that same night, team members found live tree lobsters after an 80-year absence. After being nursed with a concoction of calcium and nectar by conservationists at the Melbourne Zoo, a female *D. australis* produced enough eggs for the species' ultimate recovery.

But if the question is whether *all* insects will survive the twenty-first century, the answer undoubtedly is no. And the ecological consequences are hard to predict. In his book *Insect Diversity Conservation*, Michael Samways highlights the fact that global climate change—the grandest of humanity's impact—"involves multiple stressors and synergisms." Forecasting future outcomes is difficult, he concludes, but one thing is certain: "The insect diversity that we are encountering today at any one site will not be the same for our grandchildren."

Some insects disappear with little trace evidence as to why. And all that's left of them? Their stories.

In 1875, a Plattsmouth, Nebraska, telegrapher messaged nearby towns to verify and gauge a dark cloud 1,800 miles long and 110 miles wide composed of 10 billion locusts. To this day, it remains the largest locust swarm on record. This pest known as the Rocky Mountain locust was widespread in the Midwest. It caused $200 million in crop damage and was rumored to derail locomotive wheels as well as western expansion plans. Yet less than 30 years later, the abundant Rocky Mountain locust went extinct; the last individual locust collected was in 1902.

Mysterious as their vanishing act may be, the root obviously lies with humans. Abundance of this kind of locust could not combat the national programs for their extermination.* Bounties were put on their heads. A bushel of dead locusts got you a dollar. A bushel full of egg pods got you five. But as Jeffrey Lockwood said in a 2001 *American Entomologist* paper, "The conservationist's argument that we must save a species because we need their ecosystem services is feeble." My understanding is that he wants to use the locust's story to illustrate a larger point. Dominant as insects may be, they are not a guaranteed place-holder in a world *Homo sapiens* occupy. After all, it only took some cattle and plows to eradicate the Rocky Mountain locust. Until 1990.

Granted, the locusts were 400 years old. Using geological analyses to plot an ancient locust swarm, Lockwood found rotted specimens in a northwestern Wyoming melting glacier. They were the first collected since 1902. Were it not for global warming, writes Lockwood, the locusts may have not risen to the surface— albeit this time dead. "A century ago, human alterations of the environment caused the demise of the Rocky Mountain locust; and today, the ghosts of these insects warn us of an even more serious threat to the natural world," Lockwood said. The adaptive insect design that has survived for millions of years is being put to the test on an unprecedented level.

It's a global threat entomologists like Amber Partridge strive to prevent.

In the Butterfly Pavilion's humidity-controlled Rearing Room are various caged species getting lucky: green leaf beetles, man-faced bugs, Macleay's Spectre stick insects wobbling in a rocky sway, Red List-ed tarantulas, *Simandoa* cave roaches now extinct in the wild. The fecundity produces a damp, sodden atmosphere.

This go-around, Amber starts off with female Rosie no. 119 joined by male GRG2, short for *Grammostola rosea* and the alpha-

---

*In West Africa, locals perform a "scapegoat ceremony" to ward locusts off. It entails selecting a town member—I imagine the least popular—who is then decorated and given gifts right before being banished forever in hopes that locust swarms will follow. Returning is punishable by death.

numeric name G2. Also, I'm holding the conductor's baton—a size 12 paintbrush I'll use to tempt the pair. "What you're going to do," says our matchmaker standing beside me, "is take the brush and rub the palps like this." My hand quivers as the bristles approach G2's outstretched pedipalps. As you'll recall, these hairy straws store their semen, sucked up by the tip of their thorny embolus. My brush, frayed from years of coaxing, feels light against his pedipalp, which falls limp after each stroke. It feels like petting a very indifferent cat. I warily repeat the gestures back and forth, tickling the male and female with pheromones.

"Now you can start moving him toward her a little bit before she tries to climb out of the cage," says Amber, tone as calm and cautious as a DMV instructor. This is difficult. They barely budge. Prod one way, they go the other. "So," she says slowly, "guide him back towards her." I manage to steer them to opposing sides of the cage. As Amber retrieves another male, I continue. The spiders flinch with each rub. I try to imagine if I'd appreciate being touched like this. How consensual is this? Meanwhile, jeers come from a box of hissing cockroaches on the Rearing Room table.

There is a danger component, so I must be vigilant. "I've had some pretty bad injuries where she's gotten her fangs into him," Amber tells me. "She'll inject venom if she hasn't eaten in a while." She notices they're in the same position. "Ugh, they're so teenager-ish . . . Dude, you know what you have to do," she berates G2.

I relinquish the conductor's baton to the matchmaker. Eventually, we call quits on the Rosies and grab a pair of curly-haired tarantulas. Almost immediately the male jumps her from behind—"Sneak attack!"—and then proceeds to clumsily tap dance on her head. Believe it or not, this is a good thing. "Oh my god, this is amazing!" says Amber. "We've all seen these guys at the club. Just sayin'."

The moment passes. "She wasn't rejecting you, she didn't understand." But the damage is seemingly done. Amber returns him to face forward, stickhandling them like stubborn hockey pucks.

And then—a grand slam! Or rather intense flailing as the male makes a wrong move while crawling beneath the female. There's a quick struggle with the female wrestling the male and I jump back and scream, "Oh, shit!" My presence killed him, I think. But the alarm dies down as they face and seek out each other with intertwining legs in a Greco-Roman grapple. He drums his pedipalps on her underside, and sticks his embolus into her epigynum, filling it like a gas nozzle fills a tank.

She gets backed into the cage's corner and bares her two very ominous fangs. "She's getting mad." And possibly a little hungry. Amber breaks up the quarrel using the paintbrush. The mating was a success.

Next, data collection. Date, humidity, start/end time, and attempts get logged as well as comments. "New sperm web + short copulation" or an "A+" if sex goes smoothly. Occasionally a male will break off his embolus within a female's copulatory cavity to prevent other males from reproducing. I'm a little surprised how rapidly it all happens after the hours put in. "That's it?"

"That's it," she tells me. Except now the potentially gravid female gets a name. The matchmaker bestows the honor on me. Thinking about the surprising celebrity spider fanatic/international model I learned about in chapter 1, I name her "Claudia."

If Claudia ends up being gravid, she'll produce an egg sac that Amber will delicately confiscate and incubate herself. Egg sacs can contain 1,000 spiderlings, four of which can fit on a thumbtack. Bug reproduction is bountiful and strange, and the variety makes the process truly impressive, especially given bugs' proclivity for adaptation. How warm those breeding environments are can either damage the insects or generate swaths of breeding grounds. Mix in human travel, and we get our contentious relationship with pests and changes that have reshaped history. And it's something we'll have to face up to if we plan to live in their world.

Four

The On-Flying Things

By August 29, 1793, a month had gone by since yellow fever broke out in Philadelphia. Sparse gunfire echoed through forlorn streets as carts full of bodies were wheeled about. That's when the lively liquid contents of the public water barrels caught the eye of a person known to history only as A.B. "Whoever will take the trouble to examine their rainwater tubs," A.B. wrote in *Dunlap's American Daily Advertiser*, "will find millions of the mosquitoes fishing about the water with great agility."

Breeding conditions favored *Aedes aegypti*, and yellow fever crushed the former US capital. Over a four-day period in October 1793 alone, 386 people died, the toll reaching 5,000 by year's end. Ultimately, nearly half of the city's 50,000 inhabitants left, but the significance of A.B.'s offhand observation of mosquito larvae would not be understood for another century. The war being waged—though the concept might imply a fair fight—would make General Sun Tzu proud. "Be extremely subtle, even to the point of formlessness," he writes in *The Art of War*. "Be extremely mysterious, even to the point of soundlessness. Thereby you can be the director of the opponent's fate."

Epidemics have influenced wars, politics, and economies for millennia, as tiny insects known as vectors carry viral infections that bring devastation. The deftness of mosquitoes, fleas, and lice has stopped armies, delayed construction of grand projects, and changed the face of human civilization. During the yellow fever epidemic of 1793, George Washington's administration was temporarily dispersed. Patients filled hospitals, and the mansions that were used to house the infected were described in Jim Murphy's *An American Plague* as "great human slaughterhouse[s]." Bloodletting sometimes occurred up to 150 or so times a day. Containers filled so rapidly the phlebotomies moved outside to the cobblestone roads. The uninfected had little else to resort to but folk medicine, huffing vinegar-soaked hankies and burning gunpowder to "purify the air." This was only the first outbreak in the United States. There would be many more.

Countries that took part in the slave trade, and thus transported mosquito vectors, were the ones most likely to experience yellow fever. The West Indies were hit in 1648. As slavery in the United States increased twofold in the eighteenth century, foreign mosquitoes and an expanding population primed the country for the viral storm. Jim Murphy helps put the casualties that stretched from Boston to Savannah in perspective. Over the course of three outbreaks starting in 1853, New Orleans and Memphis reached a total of 16,000 dead. Rioters on Staten Island burned a quarantine hospital that had treated infected patients to the ground in 1858. By 1905, the outbreaks were over, shortly after their cause was discovered.

Following the Philadelphia epidemic, construction on the new US capital began on cheap, undesirable swampland between Virginia and Maryland. A breeding ground for mosquitoes, the gaseous land was associated with disease. *Malaria* actually means "bad air" in Italian. "Washington, DC exists in part today because of malaria," says US Army research entomologist Mike Turell on a phone call from Fort Detrick. "If you lived around a swamp, you got malaria. It did not take a rocket scientist to figure that out back in the 1500s." Turell, now retired, has studied arthropod-borne viruses, aka arboviruses, for nearly 40 years. When newspapers need a scoop for the next headline-fetching epidemic, Turell is the man they call.

"When a disease first happens, people get really excited about it," he explains. "West Nile virus back in late 1999 peaked across the country in 2003. There was not a *day* that there wasn't an article in virtually every single paper in the country on West Nile." Flip on the news, and you might get a clear depiction of how adults view insects. Chances are you'll see one of those superimposed nightly news teasers tucked in your TV screen's corner—something like an outline of Florida overtaken by an insect of Godzilla-like proportions. Some may feel the imagery is warranted, given that, according to the World Health Organization, yellow fever alone—protocol requires international quarantine at the sign of an outbreak—continues to kill 30,000 people annually.

Turell has traveled from Thailand to Uzbekistan netting mosquitoes and investigating potential viruses. He knows the impact an insect epidemic can have in war. "The Russians never beat Napoleon! Typhus beat Napoleon," he emphasizes. One author during World War I wrote that "many officers fear lice more than they fear bullets." Even as recently as the Vietnam War, says Turell, more soldiers lost duty time to malaria or dengue than to the Vietcong.

Nature wins most battles. Sometimes the outcomes are simply strange. "Some epidemic diseases," writes Hans Zinsser, author of *Rats, Lice and History*, "converted [the world] from uncontrolled savagery into states of relatively mild domestication." The Old English word *onflyge*, or "the on-flying things," describes these sorts of epidemic diseases. It comes from an age when a theory existed that demons caused plagues. "With the increasing concentration of the human population in agricultural communities and cities, the way was open for epidemics," write entomologists Cornelius Philip and Lloyd Rozeboom. "With no knowledge of the etiology of these diseases, it was inevitable that such catastrophes would be attributed to the machinations of malicious devils or a vindictive god." If you were to read the firsthand accounts of past epidemics, you might've believed it too.

Though its cause remains a mystery, ancient Greek historian and philosopher Thucydides' descriptive telling of the Plague of Athens back in 430 BCE was accurate enough for today's scientists to at least theorize as to what killed one-fourth (approximately 100,000) of the city's residents over a three-year period. The symptoms and consequences, which often meant amputation of fingers and toes, were noted as such:

> People in good health were all of a sudden attacked by violent heats in the head, and redness and inflammation in the eyes, the . . . throat or tongue [became] bloody and [emitted] an unnatural and fetid breath . . . When [the symptoms] fixed in the stomach, it upset it; and discharges of bile of every kind named by physicians ensued . . . In most cases also an ineffectual retching followed, producing violent spasms . . . If they passed this stage, and the disease descended further into the bowels, inducing a violent ulceration there accompanied by severe diarrhea, this brought on weakness which was generally fatal.

The Peloponnesian War began a year prior to the outbreak. Iron-willed Spartans had laid siege to Athens. So Thucydides, who'd survived the disease himself, wanted to entrust his observations to others if the disease "should ever break out again." Yet the damage to the Golden Age of Greece was done, as not only Pericles, a virtuous statesman, was killed by the plague, but as family fortunes were squandered in haste as the world ended around them. "Soldiers have rarely won wars," writes Hans Zinsser. "They more often mop up after the barrage of epidemics."

A popular belief is that the Plague of Athens can be contributed to flea-borne typhus. By then, fleas had evolved from attaching their cocoons along the hairs of our very naked primate ancestors to the fibers of our clothes, Zinsser writes, "thereby gaining a degree of protection from direct attack and a greater motility." In 2000, a research team found evidence of the typhoid fever bacterium in dental fossils from a mass burial pit in the Kerameikos cemetery, which dates to 430 BCE. (Although other scientists argue the discovered DNA fragments resemble *Salmonella* species.) The orthodontist responsible for the initial find of the typhus relative continues to conduct DNA tests on a larger variety of teeth samples to affirm his theory. If typhus was not the mystery cause of the Plague of Athens, researchers point to other arboviral diseases and bubonic plague as potential sources. Such arboviral diseases had made their woeful rounds for some time.

Officials in ancient Persia, according to *History of Entomology*, noticed the deadly "stranger's disease" at the inns travelers would frequent. While the Bible does tell the story of Moses raining down some awesome locusts, there's also a historical account of Assyrian king Sennacherib in the text. While trying to invade Jerusalem in 701 BCE, his army was smitten by a disease carried by fleas. It's possible mosquitoes even played an earlier role in shaping human civilization. As far back as 500 BCE, Herodotus witnessed fishermen using mosquito nets at night. The famous Greek scientist also believed he could avoid mosquito bites by sleeping in tall towers.* One contending theory for Alexander the Great's demise

---

*Never mind bugs' ability to hit incredible altitudes. In 1926, US Bureau of Entomology man P. A. Glick hooked a sticky trap to a monoplane to determine the migration pattern of insect pests miles above land—what he deemed "the 'plankton' of the air." On those first initial flights he found thousands of bugs at 2,000 feet and even a lone orb weaver spider as high as 15,000 feet.

in 323 BCE, courtesy of an especially observant epidemiologist, was that the West Nile virus (transmitted by mosquitoes) was at play, as evidenced by the "omen" when potentially infected ravens fell at the conqueror's feet. Insects during that time directed fates like army generals, especially flea-infested rats. During the Roman civil war of 88 BCE, epidemic disease killed 17,000 of Octavius's soldiers. A plague likewise stopped the Huns' assault on a very vulnerable Constantinople in 425 CE.

The Plague of Justinian—the first record of the bubonic infection *Yersinia pestis*—gave the "coup de grâce" to the Roman Empire, writes Zinsser, killing 50 million people in a 200-year period. It began at a Mediterranean port in Alexandria in 542 CE. No one paid any mind to the black rats on the docks, but soon more sailors reported symptoms. Travel continued and the bubonic plague followed despite the "magical amulets and rings" used to ward it off, writes William Rosen, author of *Justinian's Flea*. Death set in 17 days after infection. People began carrying name tags should they collapse and die in the street. "Constantinople," Rosen says, "was a window onto Hell."

And the Renaissance? Thank the same fleas that transmitted *Yersinia pestis*, causing the Black Death pandemic of 1346–1353. As a result, one-third of Europe's population died, which in turn propagated a number of cultural and socioeconomic changes, and wealthy patrons who supported the arts, which led to a cultural rebirth and the creative geniuses that sprang from it. Two centuries later, in 1530, when Charles V of Spain was prepared to surrender Italy to the French, a typhus plague struck 25,000 French soldiers down, handing Charles V the crown to the Roman Empire. During France's Saint-Domingue expedition to conquer what is now Haiti, Napoleon Bonaparte lost 23,000 men to yellow fever by 1803. This resulted in the Louisiana Purchase. Napoleon's luck didn't end there. From June to September 1812, his army was reduced from 500,000 to 100,000 men due to typhus as they marched to invade Moscow. Only 40,000 soldiers remained by the time they made it back to the Polish border. One-fourth of the deaths during the US Civil War (approximately 155,000) were caused by flies spreading "Camp Fever," otherwise known as typhus. And before the Civil War began, the Third Pandemic took hold of China in 1855, killing an estimated 12 million. The bubonic plague officially ended in 1959, but not before hitting India in the late nineteenth century, taking another 6 million with it.

By the end of the nineteenth century, scientists were determined to finally crack the insect epidemic code. France sent thousands of workers and engineers to build the Panama Canal. But over an eight-year period during the 1880s, over 30,000 of them died from malaria and yellow fever. The tropics became known as the "white man's grave." Panic from the preceding years of US outbreaks had peaked. ALL MAIL FUMIGATED WITH FORMALDEHYDE had been printed on the majority of letters coming from the South. The last straw came in 1898, when 3,000 US soldiers succumbed to typhoid and yellow fever during the Spanish-American War. Papers at the time mocked the idea of mosquitoes transmitting disease, so the US Army surgeon general assembled a team to find out how yellow fever spread.

In the past, doctors had tried their hand at self-infection to understand the virus, even to the point of ingesting pills made of the black vomit from infected patients.* Others thought that yellow fever may be transmitted to telegraph operators via wires and "air-electricity," as suggested in the 1881 paper entitled "The Electro-Galvanic Theory of Yellow Fever—Disturbed Electricity the Exciting Cause." Enough said. Coincidentally, that same volume of the *New Orleans Medical and Surgical Journal* contains a paper by Carlos Finlay that outlines the *exact* agent of the transmission.

---

*Dr. Luke Pryor Blackburn—a psychopath embittered by the prospect of the North winning the Civil War—was a prototype bioweapon attacker. In 1864, the good doctor, later known as the "Yellow Fever Fiend" or, my favorite, "Dr. Black Vomit," mailed clothes collected from his yellow fever patients in Bermuda to Northern cities to infect whoever opened the packages.

"It seems natural," Finlay writes, "that this agent could be found in that class of insects which, by penetrating into the interior of the blood vessels, could suck up the blood together with any infecting particles contained therein, and carry the same from the diseased to the healthy."

Finlay spoke rapidly and stuttered. It made him appear all the more kooky. "He was dubbed 'Mosquito Man' by the US press," writes Molly Caldwell Crosby in her incredibly researched book *The American Plague*. "[He] became known as a 'crank' and a 'crazy old man' in Havana." Public acceptance would take time. Finding the proof of that connection would take a curious physician at the dawn of the twentieth century whose lab methods verged on human sacrifice. His nemesis was what might be considered the principal buggy villain of the past two centuries: the mosquito.

You have to appreciate the ethical dilemmas of US Army physician Walter Reed in Cuba during the Yellow Fever Commission of 1900. At first, Reed believed a bacteria was at play. However, a young bacteriologist by the name of Jesse Lazear, who had also been assigned to the commission, was keen on Carlos Finlay's initially dismissed mosquito hypothesis. Two years before the yellow fever experiments in Cuba, scientists discovered that the bubonic plague–causing bacillus was carried by fleas and that mosquitoes transferred malarial parasites to birds. So Lazear bred mosquitoes for dissection in jars filled with bananas. But Reed saw that although the lab work would yield conclusive evidence, it wouldn't be enough to prove mosquitoes are vectors of yellow fever. A direct correlation was necessary. So the commission members enlisted a few brave men and women.

If you were asked to dip your hand in a jar full of yellow-fever-infected mosquitoes, I would hope you'd say no. But, fortunately for science, one group of human volunteers said yes. In the first experiment with informed consent of death, the subjects agreed to be repeatedly bitten by mosquitoes, specifically the female *Aedes aegypti*—they're the bloodsuckers.

The expected symptoms include searing headaches, sensitivity to light, and yellow-tinged skin. During the fall of 1900, patients at Havana's Las Animas hospital were already in the throes of the disease. Here, Jesse Lazear let Finlay's mosquitoes feed on the sick and then healthy volunteers, which included all of the scientists on the commission. Of those who were bitten, two men became sick but later recovered. Jesse Lazear, expecting his second child, continued experimenting as the anonymous "Guinea Pig No. 1" in his ledger. He died seven days after being bitten. Walter Reed was overwhelmed with guilt because he had not been in Cuba at the time of his colleagues' self-experimentation. But together the medical doctors of the commission protected Lazear, writes Molly Caldwell Crosby, so his family wouldn't be denied life insurance due to his "medical suicide."

Many involved with the commission had bitter memories of the suffering caused by yellow fever. It was especially personal for Lena Angevine Warner, the chief nurse serving under Reed in Havana. During the Memphis epidemic in 1878, as a young girl, she had lain on her house floor immobilized, sickened by the disease, and watched her siblings and guests and the family servants die. She was alive still as robbers broke into her home and strangled her father.

Daunted by the loss of Jesse Lazear, Reed presented the commission's findings and convinced the US government to fund a new facility. The new experimental ground was called Camp Lazear. There, Reed would rule out other means of transmission. One method aimed to dismiss yellow fever as a germ. At first, the "Infected Clothing Building"—a dark, sealed room filled with diarrhea- and vomit-sodden clothes and beddings—sounds as harmless as those red biohazard boxes in a doctor's examination room. But when you stick people in there, crank the temperature up to 100 degrees Fahrenheit, and keep the door locked for three weeks, the "Infected Clothing Building" seems like a misnomer. At the beginning, volunteers ran outside gagging from the smell percolating from the infested wooden crates. But you'll be happy to know they survived unharmed—at least physically.

In Building No. 2, infected mosquitoes shared space with sick and healthy volunteers in beds partitioned by cheesecloth. Another room had no separation. Amazingly, the soldiers involved declined monetary compensation as they nobly thought of the thousands of lives they'd save. They continued receiving bites through December 1900. "My spine felt twisted and my head swollen and my eyes felt as if they would pop out of my head," said one volunteer. "Even the ends of my fingers felt as though they would snap off." By January 1901, Walter Reed had confidently established yellow fever's propagation, and he believed survivors could eventually become immune to the virus.

Experimentation continued under direction of Carlos Finlay and another doctor at the Havana hospital. It didn't conclude until public protests sought an end to it after three more people died. Their families were compensated $100 each.

The moral burden of the Yellow Fever Commission is said to have weakened Walter Reed's immune system. He passed away two years later, in 1902. But the success of the commission changed the world; the first human virus was discovered, and it was confirmed that mosquitoes become pathogenic 12 to 20 days after biting an infected individual. After the discovery, Colonel William Gorgas of the Medical Corps rallied "mosquito hunters" to blitzkrieg *Aedes* breeding grounds (gutters, water pots) in 1901 and put up window screens across Havana neighborhoods. Because of the bug exterminators, only one city resident died from yellow fever over a four-month period. Word of these tactics spread and were also implemented in Central America, allowing for the completion of the Panama Canal.

Reed's legacy was solidified by the Walter Reed Army Institute of Research. Infectious disease research continues there in hopes of fighting off the largely endemic dengue carrier *A. aegypti* and the sandfly parasites US troops encounter in the Middle East. Across the pond in underground labs at the London School of Hygiene and Tropical Medicine, volunteers still shove their arms into containers with mosquitoes to test repellants. Archaic? Maybe—but we've learned that defeating big scourges requires equal parts heresy and obsession. Consider the fact that Swiss chemist Paul Müller mixed 349 concoctions and took four years of research before finding *the* Insecticide—warranting a capital "I"—known as DDT. One of Müller's drives was combating the lice-spread typhus epidemic in Russia that wreaked havoc from 1870 to 1940 and killed millions. Until the EPA banned the synthetic in the 1990s for the ecological havoc it wreaked, DDT became the go-to for defeating bed bug epidemics, plagues, malaria, and yellow fever outbreaks. (It's also been theorized that our TV-watching habits kept us indoors, thus reducing insect interaction.) Strategic use of the compound, such as with air strikes, made the world livable and bug wars winnable. But overexposure rendered many insects resistant—an evolutionary knack bugs have, if you remember from chapter 3, in order to survive in volume.

In the wake of Reed's work, in 1937, virologist Max Theiler developed "17D"—a vaccine against yellow fever that the World Health Organization concluded in 2013 gave "lifelong immunity." Shortly following the vaccine's creation, a malaria epidemic—caused by a complex multicellular parasite—hit the Americas. One of the top villains behind it? *Anopheles gambiae*—the most adept African malaria vector, discovered in Brazil by entomologist Raymond Shannon. The mosquito was eradicated shortly before World War II with a highly toxic insecticide called Paris green, applications of which were carried out by Fred Soper, a man Malcolm Gladwell called the "General Patton of entomology." Soper would go on to form the Global Malaria Eradication Program. According to the Centers for Disease Control (CDC), all traces of the parasite in the United States were "eliminated" by the early 1950s.

Sometimes, though, control methods target the incorrect mosquito, making the actual vector more prevalent. This is what happened once during World War II in the South Pacific with two different species of *Anopheles*. One of these species enjoyed laying eggs on water in shaded areas, so the solution was to hack away the brush.

"A few weeks later a different *Anopheles* species came in—one that *liked* sunlit water," says our arbovirus expert Mike Turell. That particular species had previously been rare. "The first specie was an inefficient transmitter of malaria. However, the second one was a *much* better transmitter. By controlling the wrong species, they actually made the disease much worse."

Today, the pathogens carried by *Anopheles* cause 1 million deaths per year. In 2010, the CDC reported 219 million cases of malaria—a parasitic disease with flu-like symptoms—across the world. Some years that number soared as high as 500 million. Normally *Anopheles* resides in tropical climates, but seafarers changed that. According to Michael Specter's "Mosquito Solution" article in the *New Yorker*, "Researchers estimate that mosquitoes have been responsible for half the deaths in human history." As the toxic control methods of yesteryear are outlawed, research teams look to eliminate malaria-transmitting mosquitoes and other vectors using cutting-edge sterilization and genetic modification programs.

In the battle to reduce cases of malaria and dengue fever, several solutions lie at the forefront. Two come from North Carolina State University and UC Irvine, and involve lab-raised mosquitoes. Molecular geneticist Anthony James at UC Irvine told one reporter there are two genetic modification strategies: a "bite" method that stops diseases from being transmitted, or a "no-bite" goal that eliminates certain species entirely. The latter is being attempted at NC State with the engineering of flightless females as a means to decrease the *A. aegypti* species. As for the "bite" method in Irvine, James is taking malaria-destroying genes found in mice and modifying them for mosquitoes so the blood-borne disease will be destroyed in their own bodies. (Parasites are actually stored in the mosquitoes' salivary glands, which strengthens the pathogen.) After working on it for 15 years, he made a recent breakthrough with the help of colleagues at UC San Diego. Once James's mosquitoes became resistant to malarial parasites, he needed a gene drive to make the resistance inheritable, transferring the majority (rather than half) of those altered genes to their mosquito offspring. Fortunately, a team at UC San Diego, which had accomplished similar work with gene drives in fruit flies, found out about James's efforts. By July 2015, the team at UC Irvine produced mosquito larvae with red eyes—a genetic marker that the gene drive had delivered the malaria parasite antibodies. One medical entomologist quoted in a *Nature* piece argued that "the elimination of *Anopheles* would be very significant for mankind."

As with DDT, mutations in malaria-resistant mosquitoes may occur within the gene drive over time. However, I wonder if such genetically modified (GM) *Anopheles* could be released periodically should malaria cases rise again.

It's not pure conjecture. One England-based company has stirred the gene pools in the Cayman Islands since 2010, reducing the population of harmful mosquitoes by 96 percent. Oxitec labs breed GM males of Walter Reed's favorite: *A. aegypti*. These males produce sterile progeny. Even better, the population suppression can be reversed if needed by stopping the release of GM mosquitoes over a long period of time. Why anyone would want such a thing, after considering the facts about GM mosquitoes, is hard to say. Yet Oxitec's proposal to release their mosquitoes in the United States has come under fire within the past decade, sparking controversy and debate—specifically in the prospective release zone of Key Haven, Florida. At the time of this writing, Oxitec awaits FDA approval to run a trial in areas prone to a dengue and Zika outbreak.

Fortunately, after a couple of e-mail exchanges with Oxitec scientists (read: persistent nagging), I was able to arrange a firsthand look at their "elegant approach" of combating *A. aegypti* in the city of Piracicaba in São Paulo state.*

It's 6:30 a.m. at the São Paulo airport, and the woman at the rental car company has a mascara-smudged, hungover gaze. Bits of her leopard-print dress poke out from beneath her red Avis business suit as though she were a club-going Superwoman ready to dance in a heartbeat (or bass beat). She sits in a kiosk eating cubes of wet cheese from a container, repeatedly reminding me I didn't buy insurance. Part of me foolishly disregards this, and I chalk it up to my nonexistent Portuguese tongue and our conversation becoming lost in translation.

Part of me is also beat from the nine-hour flight. As I was landing, a clear view of the green mountains surrounding São Paulo appeared from the clouds. A blanket of fog smothering the slopes reminded me of the chemical plumes used to kill mosquitoes here—a crude method that is gradually being replaced by the twenty-first-century innovation I've come to see.

I drive onto the BR-116 federal highway. Traffic in Brazil operates with the same gusto as their spoken language: full of ardent rhythm. A calvary of mopeds buzz by with quickened beeps as they cut through lanes. Tiny cars merge with zipper-teeth precision. And part of me really hopes the Avis woman was wrong as I come to an abrupt halt to avoid being pancaked by two commuter

---

* This was of course followed by a lengthy rebuke from my mother about long-distance travel: "Go to Puerto Rico—follow some mosquitoes around there! Listen to your mother. You're driving me crazy!"

buses. What really catches my attention, though, is the Tietê River running along the highway. As Brazil's most polluted river, it features Styrofoam cups, Carrefour grocery bags, and slums directly on the riverbanks, their detritus scattered about. All of it makes for a cozy home for *Aedes aegypti*. Deforestation and elimination of would-be breeding grounds exacerbates issues with dengue and Zika viruses (which causes brain abnormalities in fetuses). During the first half of 2015, 760,000 confirmed cases of dengue fever were reported by the Brazilian Ministry of Health. Over 200 people died. Reported cases tripled in São Paulo alone in the course of a year, and this was before Zika outbreaks hit hard in 2015.

"When I first came here, I was a little surprised to see dengue on the television," says Glen Slade, Oxitec's head of business development. He oversees the Campinas subsidiary in their Brazil office. A Brit, Slade's visibly pleased with the 85-degree weather. "You'll see TV adverts at night asking not to leave water in your flower pots, trays . . ." he goes on. Currently it's the dry season. But in a month, the rains will come, followed by a high concentration of mosquitoes.

"We are substituting quite a large scale of chemicals in Brazil designed to kill mosquitoes," Slade says about Oxitec. After several trial releases, aside from the ones on the Cayman Islands, the substitution of using their GM mosquitoes, aka OX513A, has proven effective. Exhibit A: the Iterberaba neighborhood of Juazerio, Brazil, home to 908 residents. Working with the University of São Paulo, OX513A helped suppress *A. aegypti* within 13 acres by 94 percent from May 2011 to October 2012. Exhibit B: Jacobina, Brazil. Starting in July 2013, their "organic product" reduced the dengue vector by 92 percent. And now Exhibit C: Piracicaba, a city only an hour's drive away from their offices here in Campinas where they've been releasing about 800,000 GM male mosquitoes a week. Annually, that's about 40 million.

Cast ye fears aside, GMO watchdogs. Not only has the National Biosafety Committee in Brazil approved the use of OX513A, but the UK's House of Lords has reported GM mosquitoes could "save countless lives worldwide." What's unique about the trial is that Oxitec is working directly with the Piracicaba municipality rather than third parties. This makes fine-tuning processes in treating the tiny Piracicaban neighborhood of 5,000 residents more concise. (The residents all receive door-to-door counseling on the *mosquito do bem*. In Portuguese, that translates to "the good mosquito.") Later Oxitec will expand to producing hundreds of millions of OX513A to combat dengue fever, which endangers 40 percent of the world, as well as the explosive outbreak of Zika virus.

OX513A is the brainchild of chief scientist Luke Alphey. What sets it apart are two altered genes in *A. aegypti* males. Remember, they don't bite. Rather than hitting these delicate, mustachioed bugs with radiation via the decades-old sterilized-insect technique, Alphey needed to keep their physical endurance intact enough to compete with wild males while breeding. So he developed a "lethal system" specific to males. One autocidal gene ensures the progeny have a decreased life span. To keep the lab-bred males alive long enough to mate with females, this gene is subdued by an antibiotic.

"When we add tetracycline, it turns the gene off," Karla Tepedino tells me. As Oxitec's Brazil production and field trial supervisor, she oversees the factory line here from egg to promiscuous adulthood. She's in her mid-twenties and wears stylish earlobe plugs, and the job has fortified her patience. "Someone always brings up *Jurassic Park*," she says, hands in the air, dumbfounded. Campfire theorists argue that Oxitec mosquitoes—despite being largely unchanged for millions of years—could rapidly evolve to carry the most potent strain of the dengue virus. "Why do you think we're going to be capable of such a thing?" Karla asks. This is frustratingly accompanied by GMO jabs and people accusing her and Oxitec of "playing god."

Genetic clock ticking, Alphey needed a marker to enable Oxitec to track their success from OX513A eggs collected in the wild. Hence the fluorescent dotted pattern, detectable under a microscope, in their larval progeny. "It's a shame they don't fluoresce naturally," says Karla, who has a plushy mosquito doll on her desk. "It would be so nice to have fluorescent mosquitoes flying around." Although these two genes are altered in their chromosomes back in the UK, OX513A must be bred locally in a controlled setting here in Campinas.

A biohazard sign on the containment room's glass door reads: *Proibida a entrada de pesos não autorizadas.* Karla Tepedino keys me in. She asks that I run my shoes through the bootie dispenser—a fun little machine you drag your foot through to wrap in blue sterile fabric (and one I'd like to cleverly reengineer back home for my socks). Like the Rearing Room back in Colorado, this sample-sized "mosquito factory" has that familiar buggy scent tied to such places. Karla pins the smell to the fish food used to feed mosquito larvae.*

My guide takes me through rows of white cages with screen-door meshes. This is where the process begins. Close inspection of cage C28 reveals a 3:1 female to male ratio of 16,000 mosquitoes. Lamb blood, served in these penetrable thin discs no larger than a Big Gulp lid, is slid into the cages to quench the females' thirst, while a sugar-saturated rod feeds the males. Eggs are laid in the bottom of the cage on thin strips of paper. Karla shows me a clear container labeled Ovos OX513A with approximately 3.3 million black eggs that have been collected from the paper strips. Dried, they resemble gunpowder and feel as potentially dangerous when ignited by water—were they not GM. In this dried condition, eggs can survive for up to a year. Next, about 10,000 of them go into

---

* The room's smell has to be leaps and bounds better than in 100-degree, 95 percent humidity rooms lined top to bottom with maggots gnawing on rotting meat. In the 1950s, the USDA sought to combat a screwworm fly epidemic eating southerners' livestock from the inside out. Doing that required the release of 50 million sterile flies per week reared on a factory level. Besides the squirming pupae, these putrid vats contained lean ground meat, bovine plasma, and a formaldehyde-water solution. Separated five-day-old pupae were dumped into canisters that were then loaded into irradiation chambers like submarine torpedoes.

water trays and hatch into floating larvae, basically two squint eyes with a tail. Each white tray is diluted with tetracycline to ensure survival and is slid into a baker's sheet pan rack to incubate at 81 degrees. In oddly sentimental ways, the process of manufacturing mosquitoes felt comparable to my early days as a pizza boy: leavening larvae in trays, prepping orders for delivery.

Karla leads me through the dough racks pillared across the room to a larvae siphon. Here the litter pass through what looks like a closed folding gate allowing males—who pupate after eight days—to rise to the surface. The main function of this factory line is to separate the females. "The first part of the process is to pass them through an LPS." This stands for larvae-pupae separator.* "Ta-da!" she says, demoing the device. "Just like a sieve. Once we are happy with that, we take them here." She presents a white bin labeled "LOT 193." The mosquitoes have graduated from grains of sand to BB shots with kicking tails, kind of like plump bass clefs. "They're sensitive to light," she says as she waves her hand over them. Her shadow scatters pupae from the surface. I give this a try. Their tiny, scurrying splashes feel like soda fizz on my palm.

She illustrates their size difference with two hand-sized trays. Though it's minute, the females are arguably larger because of the eggs they store. She asks me to guess which are males. "Is there a difference or not?" she asks, testing as a college professor would.

"I'd say these guys are smaller." I point to the wimpy lads.

"Yes. For me, it's evident."

But the possibility remains that smaller females sneak by. That's where the individual checkers come in. After sieving out pupae once more, a line of eight people sitting at a metal bench use a light

---

* A 1961 paper by Israeli biologists Micha Bar-Zeev and Rachel Galun details their *A. aegypti* LPS method by using a magnetic jar. Practical? Not so much. Larvae in their fourth stage (instar) of development were fed iron filings and were separated from the pupae when placed in a magnetic field. Those that didn't die from iron "clogging of the alimentary canal" evacuated the contents before pupation. As a kid fond of large magnets, I can't imagine how fun these bugs would be.

microscope to count the pupae with a fine paintbrush in one hand and a tally counter in the other. *Clck-clck-clck* . . . One lady goes through them rapidly. The room is full of this sound with intermittent zaps from electric flyswatters. Sample sizes of each batch are checked here. If they have over 2 females per every 1,000 males, they toss the batch and recalibrate their instruments. Finally, the pupae go into pots. Once the pupae grow into adults, workers drain the water and hundreds of thousands of male mosquitoes are stored for two to three days prior to release. We visit the holding room. OX513A float in the clear plastic containers, weightless as black dandelion seeds. "The nicest part," Karla says, smiling, "is the sound." It's a faint, ninja-like buzz. As steady, calm, and resolute as a stove hissing natural gas.

The next day, Guilherme Trivellato, Oxitec supervisor and mosquito delivery man, wakes up at 4:30 a.m. to feed his chickens. It's less about quirk than his biologist virtuosity—and fresh eggs. Like Karla, he's also young and personable with a lambent wit. By the time his helpers load 500 buckets of mosquitoes into a high-roof cargo van at Oxitec, it's 6:00 a.m., and he's brewing a pot of jolting coffee for him and me—"ugh, too strong"—that we both dilute with water. We're in the building's second story. The empty offices are gray in what light comes from the dingy fluorescent tubes overhead. I sit with Guilherme, sipping my coffee at a round table. From a nearby window I can see the van running idle in the driveway.

"We used to release in a pickup truck for a smaller project," says Guilherme. But scaling up Oxitec to treat a city for dengue and Zika will require a fleet of sturdy vans. So the cargo van has undergone a bit of MacGyver-ing. Vacuum hoses rigged to the dashboard's A/C vents run along the roof to the van's back to build air pressure. Aimed through the side window, like a miniature cannon, is a plastic tube. Attached to it with packing tape is an O-mouthed Dyson fan. When Guilherme does his thrice-weekly releases, an assistant sits in the back emptying the mosquitoes through the tube to execute swift mosquito drive-bys.

Jerry-rigged as the process may sound, Guilherme has refined it over time. Oxitec employees in the UK are developing an automatic release system. "But I like the feeling of doing it by hand," he tells me. He began working in Campinas in the fall of 2013 when their warehouse "workforce" comprised himself, "a plastic table, and a cell phone." Located within the business center, called Technopark, is a branch of Monsanto.

"It's complicated for us because what we do is completely opposed to what Monsanto does," Guilherme says. "It's genetic modification, but the aim and goal and the technique is completely different."

The van is already on its way to the village; otherwise, the mosquitoes will suffer in their buckets, as Piracicaba gets particularly hot. Daylight begins to break as Guilherme and I catch up to the other Oxitec employees. On the highway, looking out his car window, I take in the saturated green of Brazil. Along the roadsides are termite colony mounds as well as mountainous heaps of rust-colored junk and tires. It reminds me of my view of the river when I first arrived in São Paulo. "That has to be a perfect spot for mosquitoes," I tell Guilherme.

"I would *love* to work in a place like that," he says, excited by the thought. "How can you spray pesticide if the mosquitoes are a few meters underneath the pile of tires?" He sighs. "It's complicated. The public administration say it's the fault of the people who dump garbage everywhere, and it's true. And the population said that the city hall doesn't take care of a proper place to put the garbage. Well, that's also true. So, it's throwing the fault to each other, and that doesn't solve anything."

There's personal investment at play in Piracicaba. This is, I learn, the city Guilherme lived in for 10 years while studying at the university. He contracted dengue fever in 2011. "You feel nauseous when you try to drink anything," he recalls. "Pain in your joints . . .

And there's nothing you can do. It's quite frustrating." Everyone I spoke to, including my Couchsurfing host's brother, either fell ill with dengue fever or was one degree away from someone who did. When we drive into the city's Cecap barrio and visit its health center, I ask the head nurse how frequently she sees cases during the peak season. Guilherme translates for us, but even with my rubbish Portuguese I do pick up: "*Seis pessoas por dia.*" That is: "Six people per day."

As we drive, a pair of barking, grungy street dogs follows us. An older man leans on a lamppost, his rounded gut jutting from a tucked-in shirt. Guilherme lightly honks twice at him and waves. "I know everybody here in the neighborhood," he laughs. "He's the most boring guy." Someone dumps gray water from their balcony onto the sidewalk. "The city is growing a lot. And it's good to have options." He points toward the drug deal we see under way in broad daylight. "If you want to have some crack stones," he jests, "feel free. Have a cachaça* at seven a.m.? Go ahead." We follow the trail of the Oxitec van that arrived ahead of us, zigzagging through Cecap. My host parks the car. His timing is perfect as we see the van of Oxitec employees driving down the quiet street. Just as they pass, the on-flying things launch from the van and encircle me in cloud formation before dispersing. The release ratio is ten OX513A for every wild *A. aegypti*. To monitor their success, Oxitec employees place ovitraps—water buckets with fibrous wood paddles—throughout the trial zones; this is where eggs are laid. Guilherme points to an ovitrap shaded by a recessed wall. Later a biologist will collect it and stick the eggs under a fluorescence microscope to verify that their altered genes are passed to their now-sterile progeny.

---

* As one who enjoys his whiskey, I highly recommend a complex, oak-barrel blend of this rum. Specifically ESALQ-USP's Cachaça Fina, which is found in this overgrown distillery hidden in plain sight in Piracicaba's College of Agriculture. As I and my Brazilian acquaintances learned, it's very delicious. Let's just say my 700 ml bottle never made the trip home.

Guilherme and I hop in the van. After watching mosquito emancipator Augustus's technique, I'm mentally prepared to take the helm. An AlpineQuest app, opened on a tablet before me, has a GPS map outlining their release points with little dots as if we're playing PAC-MAN. Area F was rather dense with delivery blips, so Guilherme's having me take over for Area E as he drives 6 mph through the quiet village street.

"Fasten your seat belt," I'm told, to which I lift a brow, but comply. "Pay attention . . ." Guilherme's voice trails. "We are beginning . . . now."

A single beep on the map signifies we've entered a release spot. I quickly grab one of the stacked to-go soup containers filled with buzzing males. I rap it on the table, open the lid, and a black-and-white polka-dotted fleet of engineered mutants launch from the tubular airstrip. "Don't worry about tapping hard to get them all out," Guilherme advises about emptying the containers. I give one the ol' "Shave and a Haircut" knock in the tube and OX513A pour out. We continue through Area E for a few minutes, releasing more buckets. Looking through the Dyson fan's porthole, I can briefly make out a billowy squadron of bugs on a mission to save humanity.

Months after my visit, just before outbreak season, Oxitec and Piracicaba officials reported an 82 percent reduction of *A. aegypti.* Dengue cases dropped by 91 percent. And officials planned to expand Oxitec to a factory compound and hire an additional 100 hands. With a new facility and greater manpower, the company can target larger areas. Next up is a city with a population density of 300,000. What's more, the National Health Surveillance Agency of Brazil has now temporarily permitted the company to release OX513A throughout the entire country to combat the burgeoning outbreak of Zika virus.

But mosquitoes are merely one product of Oxitec's engineered insect line. Second-order epidemics carry environmental impacts that cost hundreds of billions of dollars in damage. In mid-2015 in upstate New York, Oxitec performed caged field studies of their GM diamondback moth—an agricultural pest responsible for a huge hit to brassica vegetables. While they are good news to those who grimace at broccoli, agricultural pests are an economic detriment that encourage the use of more chemical insecticides. In fact, bug-driven plagues that threaten the environments in which we live have frequently redesigned our world.

Bees, fleas, and ants engulfed ancient Mediterranean cities, evicting their inhabitants. Great Gray Dart moths ravaged Greenland during Viking times. Rice fields in southern Japan faced plant hopper outbreaks in 701 CE and again a century later. Mormon cricket plagues would have run the Mormons out of Utah in 1848 were it not for "divinely inspired" seagulls, writes entomologist James Hogue, gobbling up the crickets. Vintners were in tears during the mid-nineteenth century's Great French Wine Blight caused by the grape phylloxera. The South diversified agricultural crops in order to stay afloat due to cotton-grubbing boll weevils* that consumed 70 percent of the harvest. (A 13-foot-tall boll weevil monument in Enterprise, Alabama, aka Weevil City, signifies the town's shift toward agricultural variety.) Bugs, it turns out, trounce the land like invisible giants.

Nationally, insect pests damage 10 to 25 percent of crops annually. Invasive species like the caterpillars of the aforementioned diamondback moth deliver an annual $5 billion hit. Sap-sucking Asian citrus psyllas carry a century-old pathogen known as citrus greening disease, which causes $4.5 billion losses to Florida orange farmers. Our introduction of species produces a history a ecological mortality. Toss in our ever-mercurial climate, and the

---

*The havoc inspired musician Charley Patton to write his guitar-plucking tune "Mississippi Bo Weavil Blues" around 1908. Other epidemic-inspired hits include Charles Johnson's 1909 ragtime ditty the "Kissing Bug Rag," about America's deadly outbreak of Chagas disease, which continues to kill 20,000 annually.

epidemic outcomes can imperil habitats for centuries—even if the species are native.

Environmental colleagues regard Jesse Logan as the "Beetle Nostradamus." In 1994 the US Forest Service bug expert predicted the beetle epidemic that's since decimated over 60 million acres of North American forests. The price tag for some outbreaks can reach over $50 billion of hurt to the logging industry. Look at a sweep of Rocky Mountain or British Columbia forest, and you'll see a landscape variegation of auburn, cabernet-colored, gray, and graying conifer trees (pine, spruce, aspen, fir) touched by bark beetles. The main culprit, no larger than 5 millimeters, is *Dendroctonus ponderosae*. As they tunnel* through to the nutrient-rich inner bark of the tree, aka phloem, they disperse *Ophiostoma minus*—a fungi stored in their mouths and under their legs, which cultivates in the wood for beetles to later munch on. The fungus rots the tree faster, staining the wood with an asphyxiated bluish hue.

Outbreaks of this sort have been recorded for some time. By the eighteenth century, one bark beetle in particular was known as the flying worm. In botanist Friedrich Gmelin's 1787 *Abhandlung über die Wurmtroknis* (*Treatise on Worm Dryness*), he writes that "no pests have ever done so much harm to the woodlands as has the bark beetle." However, as one scientist found by analyzing a 250-year period of tree ring samples, such outbreaks occur every 50 years, allowing the natural character of the forest to be restored through fires fueled by a bed of dead pine needles and

---

*When females tunnel, they emit pheromones and sounds to attract males. In an effort to prevent their attack en masse, scientists at Northern Arizona University and a composer teamed up for a peculiar collaboration. David Dunn recorded beetle calls (chirps and stridulations) within trees by burrowing a meat thermometer rigged with acoustic recording gear into the trees, becoming an effective eavesdropper. Later, scientists piped in the tunes of antagonistic rival males—and a Flagstaff rock station for variety's sake—into a slab of phloem. The results in the tunes area were significant. Not only did beetles tunnel a quarter as much (0.4 cm vs. 2.1 cm per day), but laid only one egg per pair as opposed to the 204 eggs elsewhere. Next up? Broadcasting the sound—akin to a tweety bird being strangled—over entire landscapes.

trees. Once cleared, more sunlight penetrates to reinvigorate the soil. But as Jesse Logan found, changes in the climate have not only caused drought stress in trees, but aided bark beetle reproduction rates, leading to invasions 10 times larger than normal. Our environment has become grossly accommodative to native beetles.

In a 2001 paper entitled "Ghost Forests, Global Warming, and the Mountain Pine Beetle," Logan plots a series of computer models cross-referencing the reproduction of trees, beetle developmental stages, and seasonality. For his eye-opening research, Logan largely focused on whitebark pines—said to be indicators of "a healthy ecosystem"—in the peaks of Idaho's Railroad Ridge, which are generally safe from beetle outbreaks as they reside at 10,000 feet. But during a hike through the ridge's Cloud Mountains, he found whitebark pines that had been invaded and killed by beetles during the 1930s, the hottest period in the United States on record—until recently. He had found a "ghost forest"—a spectral glimpse into the damage we are seeing more and more of elsewhere today. As the average global surface temperature has risen in the past 50 years by half a degree Fahrenheit, we have extended bark beetles' reach into higher elevations and over mountain barriers. In Yellowstone National Park, bark beetles have killed 75 percent of the older whitebark pines.

In an interview with *OnEarth*, Logan was asked what can be done to save the coniferous trees. His answer was anything but encouraging: "This is a natural event on the scale of Katrina. Could you build a fan big enough to blow a hurricane back out to the ocean? The scale, the speed, is just too much." Forests have been rendered tinderboxes. According to a 2011 study, the Greater Yellowstone Ecosystem faces an increase in the frequency of natural fires from once every 100 to 300 years to once every 30 years. The rise in fires will deter the slow-growing whitebark pine. In his excellent book *Empire of the Beetle*, Andrew Nikiforuk mentions a report by the Alberta Forest Genetic Resources Council, noting "forests that evolved over thousands of years will likely disappear by 2060." All conifer forests, as reported by the Union of Concerned Scientists, "are projected to shrink by half" in the West and down to "11 percent [of their current total] by 2100."

Rob Addington, a research associate at the Colorado Forest Restoration Institute, told me that stopping the beetle outbreak is nearly impossible and that losing various tree species is "Darwinian." This idea of selection is echoed by several other scientists, including some who think beetles are perhaps preparing the land for a warmer climate. The problem, and perhaps solution, is that survival is an insect's modus operandi. USDA forest researcher Constance Millar examined limber pines in the Sierra Nevada. She found that a difference in tree genes within the same species made them more or less vulnerable to warmer climates. "Interactions among genetic and environmental factors . . . likely preconditioned [the] vulnerability" of certain areas of limber pines. Not only were some trees stress-free, but they'd also survived beetle epidemics. In her conclusion, Millar notes: "The limber pine stands we studied are likely to persist into the future, despite the heavy mortality they experienced, even with increasing temperatures and recurring droughts."

Recent figures show a decline in mountain pine beetles as they've pigged themselves out, leaving mountains of ghost forests. In Colorado, the number of affected acres peaked at 1.2 million. By 2015, the amount of affected land had dropped to only 5,000 acres. Still, the climate change issue remains. What will be planted in the stead of ghost forests is of great interest. Meanwhile the US Forest Service scrambles to create comprehensive recovery programs, spending over $320 million in the past 10 years: thinning trees to maintain healthy ones or creating a means of pheromone disruption to keep beetles at bay. In the past 15 years, over 50 bills have been introduced to Congress about what to do, with little success. In a 2014 paper by University of Montana entomologist Diana Six, she mentions that for all our efforts, "the rate of mortality of trees was reduced only marginally." That's why she's taken to studying trees that've endured beetle invasions—these veterans of war called "supertrees."

Regardless, lumber companies have found success in a shockingly aesthetic way. Demand for the reclaimed, blue-stained wood, aka "denim pine," has become its own industry. Like some macabre reminder, you can raise pints of beer around a kill pine–paneled tasting room in Boulder's Wild Woods Brewery. Shannon Von-Eschen, the proprietor of the Etsy store Twigs & Treen, carves coasters and jewelry out of local pine from these blue, tinderbox forests. Vice President Al Gore strums a beetle kill pine ukulele. Denim pine coffee tables and cabinets can sell for thousands of dollars. And in an endearingly granola fashion, people flock to carpenters at Nature's Casket for their very eco-friendly coffins.

Okay, rustic room décor aside, you're probably starting to really hate bugs again. Why not? They invade our terrain, our blood. Hell, insect epidemics carry such gravity as to restructure societies and affect how we live. Tit for tat, though. Fleas carry the Black Death, which historians argue was partially responsible for the Renaissance. Yellow fever ravaged the United States, but prompted new avenues in medical science. Every relationship has its downside. Yet sometimes, especially in our daily lives, that downside is simply too much to bear.

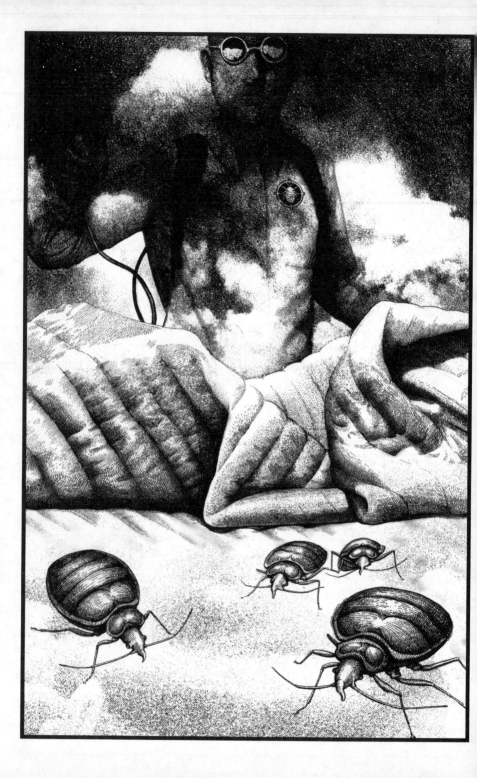

Vámonos Pest!

Cesar Soto DeLeon sips from a blue and white Anthora coffee cup while gesticulating with a pinched cigarette. "The thing with bedbugs is they don't discriminate," the pest control operator says matter-of-factly. "Number one. And number *two*, if you're well off, you're going to be in more contact with them because you travel all over the world." The two of us stand in front of a six-story apartment complex off Coney Island Avenue in Brooklyn. Cesar is here to kill. Cesar—a 50-year-old, bullet-wounded, Bluetooth-wearing Puerto Rican and reformed ex-con who grew up in "do or die" Bed-Stuy—is one of the city's unspoken finest. "You're in and out of airports. You go to Africa, go to Asia, go to Mexico—"

"And they get inside your luggage," I say.

"Exactly," he says with a prolonged emphasis, his Brooklyn-native slur rendering the pronunciation "*egg*-xactly." It becomes a bit of a catchphrase as we make our way inside to confront the terrorizers reigning supreme in apartment 5H.

We are hunting for *Cimex lectularius*, a pest Henry Miller described as "a cosmopolitan blood-sucking wingless bug of reddish brown color and vile odor," and whose PTSD-causing, suicide-inducing infestations earned the US pest control industry $470 million in 2014 alone. The United States has been in the midst of a resurgence since 2006. The "Don't Let the Bed Bugs Bite Act of 2009,"* though not passed, was introduced in Congress by Representative G. K. Butterfield in the hopes of mandating monetarily state-assisted inspections at motels. Since 2010, advocacy groups like New York vs Bed Bugs have mapped the number of bedbug complaints† by neighborhood. Citywide, there were 82 landlord violations in 2004. Looking at the map now, Bushwick appears as a bright red epicenter in 2010, with over 1,000 violations—a fourth of the total from all boroughs. That same year was deemed the "Year of the Bed Bug" by CBS, Brooke Borel notes in her book *Infested*, an account of the bedbug takeover. One artist group began placing lawn gnome–size bug hotel facades at the base of bedbug-reported buildings as a sort of hobo code to warn people. "*BEDBUGS!!!*," the off-Broadway musical featuring mutant buggies and their "glam god named Cimex," appealed to sci-fi geeks and traumatized victims while receiving positive reviews from the *New York Times* till it was "terminated." The bedbug-related revenue for Brooklyn-based Northeastern Exterminating, according to an interview in trade journal *Pest Control Technology*,‡ went from zero percent in 1994 to 76 percent in 2015.

---

* Borrowed from the "sleep tight" bedtime rhyme, which may originate from the bed's "latticework of ropes" as pointed out by bedbug historian Michael Potter. In the 1500s, straw mattresses sat atop ropes that required tightening. The mattresses were burned once they became overly infested with bugs. Thirty thousand years ago, cave dwellers—a researcher from the University of Witwatersrand theorized—burned hay beds to avoid a restless night's sleep as *C. lectularius* gravitated from bats to man.

† A range of communities report infestations, but ritzier homes tend to swiftly nip the issue in the bud. Due to the high cost of extermination (averaging $1,500 per room at times), bedbugs tend to linger. In 2016, researchers at Rutgers University reported on 2,372 low-income New Jersey apartments in four cities, including Paterson. Of the 88 residents who knew they had infestations, 57 percent of them had known for six months. Thirty-six percent had known for an entire *year*.

† I should note my interviewee had the "prestigious" honor of gracing the cover of the September 2013 issue *Pest Management Professional* magazine.

So when I ask Cesar, owner of insect-emancipating Freedom Pest Control, if it's annoying to return to a complex he's treated once before, really, it's bittersweet.

"Yes and no because I live off this," he humbly admits. "Bedbug treatment: *Ka-ching, ka-ching.* You know what I mean?"

As our population grows, so does the pest management industry. According to one analysis by the US Bureau of Labor Statistics, service revenue in 2014 was $7.5 billion, with a five-year growth rate of 3.4 percent. So I'm not surprised to hear about Orkin's 26,000-square-foot indoor training facility in Atlanta, replete with full-size house, restaurant, and supermarket simulations.

What makes Cesar stand out is how he finds "spiritual" fulfillment in killing bugs and his use of an eco-friendlier method, earning him the title of pest control operator as opposed to the more derogatory sobriquet "exterminator." Nozzle Heads. The Spray Jockeys. Rugged "therapists" to the squeamish, says entomologist Robert Snetsinger. Exterminators evoke the Wild West, or the spray gun–holstering John Goodman from *Arachnophobia*. For them, every bedbug bite is a victory flag for our enemies.

As I rifle through my, ahem, *vast* comic book collection, I'm reminded of Simon Oliver's Vertigo series *The Exterminators*, particularly the scene in which veteran spray jockey A.J. schools a newbie on the cojones required for this, our War on Bugs:

<u>A.J.</u>: You gotta realize what we *are*, Henry.

<u>Henry</u>: Pest control specialists?

<u>A.J.</u>: No, what we *really* are.

<u>A.J. (cont'd)</u>: Back to the motherfuckin' beginnin' when that caveman reached out and fuckin' squashed that first *bug* on his cave wall. It was primal instincts, *them* against *us*.

<u>A.J. (cont'd)</u>: And to make it as a bug brother you gotta get that primal bug juice on

```
        your hands, motherfucker . . . and deep
        down you've gotta get off on it.
```

While synthetic chemical pesticides dominate the marketplace, swells of exterminators opt for environmentally reasonable means outlined by the integrated pest control (IPM) government policy to "minimize risks to human health and the environment." (Look at a can of Raid Flying Insect Killer, and you'll notice one active ingredient is d-Phenothrin—a neurotoxin that causes spina bifida and at high doses hydrocephalus—convulsion, mental disability, death.) IPM techniques might include application of a 200-year-old insecticide, the siliceous sedimentary rock known as diatomaceous earth* (DE), steamers that heat bugs to death, or general house maintenance (repairing cracks, vacuuming, er, not kicking spilled Lucky Charms under the oven†). For instance, the poisonous gas known as methyl bromide, which has killed its share of exterminators, was used for decades to treat structural termite infestations—a $1.4 billion industry. Even though methyl bromide (also used on strawberries) was outlawed in 2005, sulfuryl fluoride, which aerates faster and leaves a smaller trace in wood, has been used as a viable alternative.

Still, dishonest spray jockeys swindle customers by mixing potent toxins into their spray tanks. They do work, but in the process they also create more resistance.

Cesar is a veteran bug whisperer; he started in his brother's pest control operation around the year 2000. He's met spray jockeys

---

* Mining DE, however, is hazardous enough as to cause lung cancer. DE is comprised of the "skeletal remains of diatoms" (i.e., fossilized algae), the makeup of which includes low levels of silicosis-causing crystalline silica. Researchers from the University of Washington looked at the health records of 2,342 California miners from 1942 to 1994 and came away with perturbing results. The Occupational Safety and Health Administration strives for agreeable conditions to keep carcinogenic exposure to 1 out of 1,000 men. DE miners had an increased risk of 19 out of 1,000, proving exterminators aren't the only vulnerable ones on this battlefront.

† Guilty.

who "cut" already synthetic chemicals with the off-the-shelf concentrated stuff you buy at Home Depot. This results in clients impressed by the "instant death" delivered to said buggy assailants. And while such instant gratification might be cheaper, it becomes a recurring problem, as we'll later see.

The issue with the Brooklyn apartment complex Cesar is visiting again is a stubborn elderly couple who hasn't followed his protocols. Now their opportunistic pests have trickled from their 5H apartment to residents two floors below. So, Cesar's given the landlord a reduced price. In the past he attacked bedbugs via heat treatment—one time using up to 10 heaters in Yonkers—which can cost a couple thousand dollars. For that, pest control operators (PCOs) convert a home into a sweat lodge, hiking indoor room temperatures up to 180 degrees Fahrenheit for 6 to 12 hours.

In the War on Bugs, a sick elation arises from witnessing the casualties bake on the battlefield. Of course, you run the risk of displacing bedbugs, which equates to more clients, more chemicals, and more revenue.

"Your bedbug ain't gone," adds Cesar. "You put them in this *heightened* awareness." His young assistant Orlando has closely cropped hair with stalactite sideburns. He gathers equipment from their van as Cesar finishes his cigarette. "So they'll go dormant, for lack of a better word, for two or three weeks, and then they're back." Sometimes they go without food for over a year. "They never left."

"Right," I agree. "They're just hiding in—"

"*Eeegg*-xactly. Floorboards, baseboards, outlets, curtains . . ."

Cesar, Orlando, myself, and a metal tank of Cryonite squeeze into a tight, vintage Otis elevator with a tarnished brass door. We ascend slowly, the cables sounding as though they're fraying apart. Cryonite is the polar opposite of heating, blasting liquid carbon dioxide at 108 degrees below zero through a gooseneck wand. According to an episode of YouTube's Bed Bug TV, it's especially useful in killing eggs that won't be easily affected by chemicals. But scroll through an online discussion board for PCOs about using steam versus Cryonite—if you have spare time for sporadic bickering—and the opinions vary.

We exit and walk down a quiet hallway with a patterned tile floor. When we get to 3H, Cesar tells Orlando to knock on the door. A lady with an Eastern European accent answers.

"Hello, my dear," Cesar greets her, leisurely strolling into her living room as she chases her son down. They have packed their belongings and moved everything to the center of the room, as instructed by Cesar. "Have you had these boxes here since Friday?" he asks, concerned whether the bedbugs have hidden elsewhere besides the baseboards.

"Yesterday I find it here," she says. "One. Only one."

"Okay, can you go into the hallway and wait till we come out? Then leave for an hour and come back, okay?"

"One hour?"

"One hour's fine," he says. "We'll open the windows so it can ventilate."

She switches off the *Teletubbies* and they leave. Cesar flicks on his inspection LED headlamp. He illuminates a couple of ink spot–sized stains in the closet space. This digested blood is frass, fecal evidence that he should also target this spot for bedbugs. Insect shit is the least interesting thing these inspections can turn up. Sometimes they reveal kilos of heroin under mattresses, cockroaches, and illegal firearms. Occasionally Cesar has to fan away wafts of cannabis smoke, fumigating bugs with a contact high. Some customers offer to trade in lieu of money in whatever business they do: photography, guitar lessons, etc. And then there are the solicitors.

"We've had women open up the door *buck* naked," Cesar says, making Orlando laugh. "I'm like, 'Girl, do you really think I wanna flirt with you?'"

"You probably don't want to jump into that bed," I add.

"*Egg*-xactly," he says, explaining how they'll show bites *all* over. "They go, 'Look. Look!' And she's bit right on her nipple." His Brooklyn tongue pronounces it *nibble*. "I'm like, 'Madam'"—exasperated—"'do you really think I wanna see your *nibble*?'"

According to Orlando, this can happen up to twice a month.

Back to the job at hand: the tip of the Cryonite pipe gun bends like a dental saliva ejector and makes a similar hollow noise as it moves against the crannies on the lilac wall. On the door paneling are two penguin stickers. Piñata streamers dangle off the ceiling like an indoor festival. They brush against Cesar's pipe gun as he shoots frost into the molding above the ceiling, a plume of snow falling past him. If you take the word *bug* for its ghoulish origins, Cesar Soto DeLeon is a ghostbuster. His gun outlines the corners and bases of the room and encircles family photos hung on the wall.

The same procedure is repeated in 4H and then 5H—a dwelling with a light Miami zest, a velvet painting of Venice, and, behind window curtain number one: a mass of bedbugs. "Oh, pay dirt!" He lifts a dog-ear fold of the valance to unveil a cozy orgy of dandruff-colored baby nymphs huddled with dozens of scab-flakey adults. (The physiological drive that compels them to wedge between surfaces is known as positive thigmotaxis.) "I shoulda brought a video camera to video all this shit," Cesar says, disappointed.

It's close to 5:00 p.m. and he's tired. A bandage on his stomach sweats off. Last week he had another surgery related to an old scar from a past life, before his time on Rikers Island, when he was robbed at gunpoint and shot with a .45 caliber in 1989. The bullet, which hit the area where his spleen used to be, made holes in his intestines—it was found 10 years later in his knee. "I was dying of self-contamination," he says when he talks about the surgery. But he's taken everything as a wake-up call to drop the smoking and to exercise more. The reason is inked in Spanish on his arm—the spot for his now two-year-old son which reads: *Me vida Logan*.

Today he focuses his time on the business, his family, and the talks he gives to youth groups.

Following behind him, Orlando puts down a residual layer of insecticide with a B&G spray can filled with EcoVia—a natural product with a pungent dose of thyme. Next he hits the tenants' possessions with a bottle of Cesar's own citrusy mix. To demonstrate its safety, Orlando spritzes Cesar's hands and I watch him smear the liquid on his face and mouth.* As it drips off his chin, he says, "It's EPA exempt."

But even as Cesar and Orlando lay down a residual chemical to kill any bedbugs limping away from the frost treatment, I question the methods of PCOs and wonder if this is simply an internecine war with no clear victory in sight. Efficacy versus gimmick. For instance, the product EcoVia is classified as 25b, code for "minimum risk," and thus EPA approved as the ingredients in 25b mixtures may even be harmless to insects unless sprayed directly on them. Still, EcoVia and similar products have been found to be effective if applied properly by PCOs like Cesar.

You might ask then: isn't there a "magic bullet" for all bugs? The answer has confounded entomologists and pushed chemistry and ecology into an unforeseen and complex world.

Last year, the Bed Bug Genome Consortium's champagne-popping moment came in the form of 14,220 protein-coding genes—a "blueprint" of *C. lectularius*. The *Nature Communications* paper with its 80-plus authors details the sequenced bedbug genome,

---

*A wiser choice of beverage compared to the DDT-guzzling professors and prisoners of the 1950s. Advocating lecturers famously took large gulps of DDT "to demonstrate its benign nature." If that wasn't enough, Dr. Wayland Hayes of the Centers for Disease Control enlisted US prisoners across several penitentiaries to ingest 35 milligrams of DDT for two years. No immediate effects were reported. However, one group of researchers traveled to an Italian island 50 years following a mosquito eradication program. The surveyed locals exposed to the insecticide showed an increase in liver cancer.

analyzing some of the driving forces behind traumatic insemination (recall from chapter 3), how certain bedbug lineages favor humans over bats, and, most important, the pesticide-detoxifying enzymes that paved the way for the bedbug resurgence over the last two decades.

To figure out how insecticide resistance works, let's take a quick look at one chemical commonly used on bedbugs: pyrethroids. Derived from crushed chrysanthemum flowers, synthetic pyrethroid is an axonic poison. That means it paralyzes the insect's nerve fibers. It does so by unblocking sodium channel proteins, keeping them open to flood the nervous system with overstimulation and, in the process, killing the pest. However, given insects' renowned ability to rapidly reproduce, genetic mutations occur. This "knockdown resistance" spreads so rampantly that in Australia, for example, one strain of common bedbugs was 1.4 million times more resistant to permethrin than bedbugs in Germany. Elsewhere, such resistance is "10,000-fold higher." According to the paper, the relevant mutations are V419L and L925I. Through this genome sequencing—part of Baylor College's i5k pilot project—it's now possible to find the molecular markers responsible for the resurgence.

The consortium wasn't alone in their genetic sequencing. Published congruently in the journal was a comparison map of varied bedbug strains throughout New York's five boroughs: Manhattan, Bronx, Brooklyn, Queens, and Staten Island. Whereas the strain studied by the consortium researchers was a pure form unmarred by insecticide overuse, collected first by Harold Harlan at Fort Dix in 1973, the map charted the RNA variations of the *C. lectularius* found in the 465 interconnecting Metro subway stations. The work, done in part by the American Museum of Natural History (AMNH), revealed how different insecticides made certain bedbugs of the same species evolve differently. An urban phylogeographic map, for example, traces how one Manhattan strain living in subway benches migrated to Brooklyn and branched into different strains.

"Look at people," says Lou Sorkin, an AMNH senior scientific assistant and one of the paper's coauthors. "People don't all look alike. Bedbugs look alike. Species look alike. But biologically they have different genes that are turned on or off in that population. Some [bedbugs] have a resistance issue with pyrethroids. Some have a thicker cuticle. Some have different enzymes that are turned on so they can render a pesticide application unusable on them."

That's why manufacturers enjoy testing assays on the Harlan strain of bedbugs—named after the original pest-obsessive Harold Harlan—which is susceptible to almost any poison you douse it with. But to adequately analyze them, Sorkin encourages his network of PCOs—he occasionally teaches classes to recertify them—to collect as many of the wild urban pest as possible. In fact, for a while he'd become the talking head for the bedbug craze. One article proclaimed Sorkin the "Bedbug King of New York." To wit, it's a title he's earned. He's infamous, like Harold Harlan, for feeding thousands of these pests his own blood, positioning them on his hand with a size zero paintbrush, and letting them graze like dairy cows.

"My wife was a bit trepidatious at first," Sorkin comments about releasing them on their kitchen table. "But it's fine . . . They're not trying to make a jailbreak, but if a vial drops while it's open, do they all get back in the vial?" he says with a shoulder shrug and a smile. "I don't know." Cesar Soto DeLeon also learned to feed bedbugs his own blood; like Sorkin, he raises bedbugs both to study and to sell to accredited researchers. (And not into the hands of sly PCOs who might plant them in homes, bombarding potential clients with an infestation.)

So when Cesar invited me to his Bronx apartment for a "meal," I couldn't say no. That's why I find myself in his office command center at home this evening, blowing into a Mason jar full of bedbugs. The brown wood floor matches the brown of Cesar's curtains, permitting jagged slivers of light to enter his office. The bedbugs in the jars, lined on bookshelves, might otherwise shrivel under the sun. Papers and binders are strewn about. And his wife, brother-in-law, and son Logan are in the family room watching TV. Like the rest of the apartment building, components of his kitchen are in the middle of renovation.

The carbon dioxide allure of my breath, which is what summons them to you at night, excites them. (A technique Sorkin occasionally uses to suss out bugs during home inspections is to employ a bendy straw.) Nymphs and adults crawl over a square of graph paper bespeckled with black frass. "You see the guys moving around in there?" Cesar asks me. Logan staggers into his office, playing on the floor with a toy truck that embodies Las Vegas lights. Cesar shows him the jar and asks, "What's this, Poppa? What's this?"

"*Beh-bugs*," answers Logan.

"Yeah, that's bedbugs," Cesar says, elated, turning proudly to me. "My two-year-old."

"Okay." He scoots Logan to the door. "Get your butt outta here because Daddy's gotta work. We don't want you causing another accident like you did two months ago." In that case, a very curious Logan unscrewed a jar, and Cesar got a frantic call from his wife to get home immediately.

But tonight is a controlled experiment to test my threshold for the freaky. Cesar has fed the bedbugs so often, he says, that his "burn mark"–like bites disappear after a day. With trepidation, I tip the Mason jar upside down over my inner arm. I can feel the bugs move slightly through the translucent nylon stocking thin enough for their piercing mouthparts to get through to my skin. My host goes into another room, giving me privacy. I place my recorder on the desk. Regrettably, I keep it rolling.

Listening to it later, I can't tell if that's me blowing in the jar or sighing in frustration in anticipation of the tiny pricks or in self-loathing for being this curious. A bedbug's mouthpart syringe is called a rostrum. Piercing mandibles protrude from this inverted cone-shaped device to do the sucking, while anticoagulants in the bugs' saliva travels down another tubule to keep the blood flowing. "I maybe feel a slight—*ugh*—crawling around," I blurt aloud to myself, senses heightened, "but I don't feel pinching or any-thing. Let's see . . ." I huff in dread. "I'm a little freaked out."

The rest is too embarrassing to share.

Cesar returns. After he briefly taunts me as a "chicken," your correspondent musters whatever mettle there may be left in him and lets the bedbugs bite . . . for 12 minutes. Examining the jar afterward, I notice a couple of plump, ruby adults satiated from the session.

Cesar's wife Maria pops in to say hello. "Oh, my god, you're feeding bedbugs—ooooh, brave you! *Buh*-raaaave you."

"Have you ever done it before?" I ask.

"I did involuntarily in my *sleep*."

Cesar grins. "She was sleeping and I put a bottle up on her. And they ate. Her. Up! She had the biggest welt." The perks of living with a PCO. Additionally, the family's pet beagle, Tre, works part-time sniffing out* bedbugs, which carry an identifiable, herbal scent that Lou Sorkin compares to "cilantro, coriander, [and] cit-ronella."

In two weeks, an inflammation appears on my arm. Just a few marks at first, gradually more. The red constellation of bites mu-tates into a galaxy. Mottled swellings condense into well-defined dots. Bedbugs tend to follow the road map of our arteries and veins. Though I now know what it feels like to be bitten, the recur-ring nightmares others have lamented will hopefully remain a mystery to me. I now understand how an industry can make bil-lions in an arms race to combat an invasion and how some exter-minators can achieve sainthood.

"May The Destroyers Of Peace Be Destroyed By Us." So pro-claimed the "gas-lit sign," writes entomologist Michael F. Potter,

---

* A bedbug-detecting dog, according to one *New York Times* article, can cost as much as $11,000.

above the shop of London's very own Tiffin & Son, "Bug-Destroyers to Her Majesty," in 1690. By the seventeenth century, nobility desired total comfort, Potter notes in his paper "The History of Bed Bug Management." The wealthy simply refused to put up with the nasties shared by their servants, who were, as Tiffin said, "apt to bring bugs in their boxes." And so came the birth of the pest management trade. At this time, kidney bean leaves placed under the bed trapped *C. lectularius* with their prickly fibers, ensnaring their legs like bear traps, while other would-be exterminators practiced foolhardy endeavors, like sprinkling gunpowder on the bed and igniting it.* A couple of decades later, Englishman John Southall— whom *Insect Lives* editors Erich Hoyt and Ted Schultz consider the "patron saint of exterminators"—began a trend that grew exponentially in the 1900s: the insecticide market. It began in 1730 with a 44-page pamphlet entitled *A Treatise of Buggs*.

Holding it now in the AMNH library research room, I have to admit I'm a little agog. First, I love the smell of books from centuries past (like an old spice cupboard). Second, printmakers still used the long *s*, which looks like an *f*, to start words like "furprized"—a surprise party with furs? And third, the language is comparatively more pompous than some of the most braggart writers of the twentieth century. John Southall does not disappoint in the retelling of how he discovered a bedbug insecticide in the West Indies in 1726 to later eliminate that "nauseous venomous insect" that had been terrorizing the sleep of Londoners for centuries. He met an old man who'd formerly been enslaved who was kind enough to share his personal solution for stopping bedbugs. Southall then stole the recipe for a few pennies, "one piece of beef, some biscuits and a bottle of beer." That beer was then replaced with the tincture by the old Jamaican himself:

---

* Add to that extremity the kerosene-doused feathers used to eliminate bedbugs into the twentieth century. This is certainly less dramatic than the occasional blowtorch taken to them today.

On my arrival at London in August 1727, I made some Liquor to compare with his, (which I found exactly the same) whereupon I set about destroying of Buggs, and found to my Satisfaction, that wherever I apply'ed it, it brought out and kill'd 'em all. At length I advertise'd, had great business, and pleased everybody, then apprehending no return of the Vermin. But yet, to my surprise, tho' I had kill'd all the old ones, young ones sometimes, in some places, would appear.

Southall sold copies of his *Treatise* to Londoners for one shilling and the nameless Jamaican's "Nonpareil Liquor" for two. The secret recipe may have been a derivative of quassia wood. "I have determined by all means possible," he wrote, "to make their destruction my Profession." Unlike Southall, his pest-killing elders, the Tiffin & Son exterminators believed that "bug poisons ain't worth much, for all depends upon the application of them." A sentiment echoed by many PCOs worried about an industry dictated and perpetuated by our fears.*

Anti-pest methods developed in England. In Henry Mayhew's *London Labour and the London Poor*, the Victorian journalist documents his 1840s run-in with "Catch-'em-Alive" sellers. Boys on the street selling flypaper—an admixture, Mayhew states, of boiled oil, resin, and turpentine—called to the passersby, crying in a singsongy way: "Fly-papers, ketch 'em all alive, the nasty flies, tormenting the baby's eyes. Who'd be fly-blow'd, by all the nasty blue-bottles, beetles, and flies?" Modern flypaper didn't "uncoil," tells one *Milwaukee Journal* article, till 1863 when German baker and inventor Frederick Kaiser was frustrated with those cake-loving pests. Originally he used wallpaper dipped in molasses,

---

*Perhaps my favorite embodiment of this is *Saturday Night Live*'s "Bug Off" skit. The commercial parody has Will Ferrell advertising a medieval cockroach torture bait trap in which tweezers split the legs in different directions and beat the roach with them, as a "red-hot, metal coil burns off its reproductive organs" and a metal arm pendulums food "out of reach."

hung by the window's cake displays. The family-operated Aeroxon business tweaked the invention for a century and a half, resulting in today's cutesy butterfly sticker traps.

Pesticidal inventions of the early 1900s moved us beyond Black Flag's Quick Loader—a tin dust blower used to apply pyrethrum powder to bedsheets. Indoor vapor treatments with coal tar–derived fumigants became popular. And like the toxins percolating in the atmosphere, US government health departments scared homeowners into daily pesticide use in the early twentieth century as news reports of typhoid spread. In her book *Pests in the City*, Dawn Day Biehler writes: "In the 1910s, health departments" promoted the use of tools and traps to "emphasize private responsibility for fly-borne diseases . . . Household fly control became one element of domestic science taught in classrooms and by in-home educators." The typhoid death rate in the United States was then 15 out of 100,000. (Down from 31 in 1900.) L. O. Howard, chief of the Bureau of Entomology from 1894 to 1927, tried a little experiment in 1908 with his employees, whom he asked to hang fly-papers in their Washington, DC, homes. Although 2,700 flies were caught, it's hard to say if the experiment affected the number of typhoid cases. During World War I, posters depicted a swarm of flies invading a home from the garbage. "Our greatest menace is domestic, not foreign!" it read. Hydrocyanic acid gas (HCN) fumigation was encouraged by government entomologists at the time. By World War II, typhoid had declined due to improved sanitization. However, by the early 1900s, we were already hooked on insecticides. "Clever advertising" played a significant part in creating a market for various weapons, writes organic farming authority Will Allen. And some of the best came from a young, imaginative artist named Theodore Geisel, better known as Dr. Seuss.

Before the days of green eggs and ham, Standard Oil sought out the cartoonist to create ads for Flit, a petroleum solvent meant for residential insecticide use. For 15 years Dr. Seuss drew up inventive depictions of the Flit spray gun's use, including a plane reminiscent of the bicycle pump design. I stumbled upon one such illustration in San Diego's Geisel Library. It depicted a dive-bombing "Bug-Buzz" airplane. In other ads, you can see the sinister mosquito-like doppelgänger of Mr. Grinch. But the success of Flit's campaign had devastating consequences.

On April 23, 1940, the use of the petroleum solvent led to the deaths of 209 nightclub goers. Sprayed over the Spanish moss decorating the Rhythm Club in Natchez, Mississippi,* methane gas from the petroleum permeated the venue. Wooden boards had been nailed over the windows to keep the affairs inside private. When the methane ignited and a fire broke out, the 700 attendees were left few options but to suffocate and trample over each other in the ensuing panic.

Death, disgust, and disease seemingly follow insects.

Cockroaches are some of the most unloved bugs. Specialists cite the diseases and bacteria, like *Salmonella*, roaches can carry, but that's more of an issue in restaurants. Hopefully. On a domestic level, roaches are symptomatic of other problems like nearby rotting food, filth, or feces. But they themselves are obsessive-compulsive about their cleanliness. (They constantly lick their dirty feet clean.) For all the disgust they evoke, cockroaches ain't half bad. If anything, rashes and respiratory issues result from allergens (tropomyosin) on their exoskeletons. But since one 15-millimeter female breeds up to 300 eggs capable of making four generations per year, each enjoying, like us, warm places with food, it's no wonder they make awful pets and excellent targets for extermination.

"Roach Chow," one of the bigger precursors to modern household pest control, actually had low toxicity. Invented by P. F. Harris in 1922, these boric acid tablets staved off cockroaches, ants, silver-

---

* In 1996, Mississippi was hit with an insecticide-related fallout affecting a wider area. A poison intended for boll weevils known as methyl parathion was illegally obtained by exterminators and applied for two years to 1,500 homes and businesses, according to one report. To avoid a complete disaster (since it had caused over 20 deaths in the past), 1,100 people were relocated.

fish, and termites by damaging their exoskeletons and attacking the cellular lining of their guts, ruining their metabolism. In 1924, to prove its effectiveness, Harris aided the White House in killing off hordes of insects. "Once they get my stuff inside them," he told a *Washington Times* reporter of his natural mineral, "they are done for." (A similar guarantee for certain motels: "Roaches check in, but they don't check out!") Harris's campaign yielded five pounds of dead insects. As of 1979, a family home in Schenectady, New York,* has the honor of having had world's largest cockroach infestation with over 1 million bugs. Exterminators for this operation filled pucks with a bait called Maxforce (you know it today as Combat), which first poisons a portion of the horde and more ex post facto via roaches feeding on toxic poop.

But what of long-term success? Contrary to what the ant-like Borgs of *Star Trek* say, resistance is *inevitable*. German cockroaches, North Carolina State University scientists discovered in 2013, are no longer "checking in" to Roach Motels. They've adapted. The once-tasty glucose used in Black Flag products, thanks to a rapidly evolved genetic mutation, now tastes bitter to cockroaches. Our search for a "magic bullet"—or bullets—has led to today's endless rotation of bait inventions and insecticide sprays. Today the monetary toll of now-resistant pests is about $60 billion per year.

Insects were able to defeat what the comic book geek in me calls the Four Horsemen of the Bugocalypse. Chemical admixtures developed initially as an agricultural means of *control* whose use transitioned into one of *annihilation*. The result was increased resistance, ecological disruption, and loss of human life. The First Horseman that marks this epoch was known as Paris green. "The rise of Paris green left a historical legacy of ambivalence to toxic insecticides," writes *Bugs and the Victorians* author J. F. M. Clark.

---

*Just don't annoy a PCO on the clock. An ornery Manhattan taxi driver made the mistake of nearly sideswiping one. Unbeknownst to the cabbie, the PCO had just finished capturing a bag full of roaches from a nearby bar, which he poured into the cab's window after a heated confrontation.

And it began with the eastward march of the voracious Colorado potato beetle.

In the mid-1800s, fields across the United States became giant salad bars for the black and yellow beetle, which is capable of laying up to 800 eggs. In their larval form, they can consume up to 40 square centimeters of foliage a day. By the 1870s, illustrated WANTED posters for the now-transatlantic beetle were created by UK customs, informing inspectors and workers that "the Insect if seen, to be crushed at once." This angered entomologists trying to get samples for proper ID-ing. In the United States, people began applying to crops an emerald paint pigment containing arsenic known as Paris green. A Michigan man reportedly first used Paris green—intended to kill rats in its eponymous city—against beetles in 1867. It was confirmed as an effective insecticide in 1871 by USDA entomologist C. V. Riley—an important figure in this war. Bureau of Entomology chief L. O. Howard also helped guide the rising trend of arsenic- and lead-based chemical control. The attraction was clear to nineteenth-century agriculture writer Frank Sempers. Tell a cabbage farmer that one ounce of powder will kill leaf-eating worms in a day or two, he says, and society's reluctance to using poison on crops diminishes. (Lead arsenate wasn't banned until 1988.) Although it initially worked on the beetles, its downsides soon came to light. The highly neurotoxic copper acetoarsenite caused a chronic "pins and needles" feeling in some of the people who applied it.*

Its use continued. A South Carolina plant disease treatment book from 1909 gives tips and recipes on different Paris green admixtures. It mentions that "burning foliage" is to be expected: "Some of these apparatuses are provided with gauges which will

---

* The advent of which goes back to C. V. Riley's nozzle. His 1884 applicator invention debuted in front of an audience of French vintners suffering from the Great French Wine Blight as a means of dispersing chemicals to fight mildew and grape phylloxera. This prototype for compressed air spray applicators outclassed old methods. Gone were the days, hopefully, of blowing arsenic-based London purple off of a folded piece of paper. German and Swiss improvements on Riley's nozzle led to the modern knapsack sprayers of the 1950s.

enable one with good judgement to apply the powder very rapidly in concentrated form." How the words "good judgement" made the cut beats me. The guidebook also mentions what might be considered our Second Horseman: hydrocyanic acid gas (HCN).

HCN, an inorganic insecticide smelling of bitter almonds, first came into use in 1886 to fight wood-plant grubbing scale insects. One of the most toxic fumigants used up to that time period, it also became a means of delousing. Later HCN was released by brands like DuPont under various names, and is still used occasionally today. (Agriculture comprises about 90 percent of the $13 billion insecticide market, which includes companies like Bayer and Dow.) It had the miraculous effect of paralyzing pests. A 1933 Cyanogas ad with a man wielding a cyanide gas backpack beckons, "Fight the hoppers! Enlist with me against the grape leaf hopper." Similar industry advertising and government-led campaigns reinforced insect hostility. It seems as though we are ages from what entomologist Julian West called the "Period of Friendly Tolerance" of houseflies before the twentieth century. HCN's standard use involved enclosing groves with a tarp and then releasing the cyanide gas. Similar enclosed use was administered for indoor bedbug fumigations—with obvious fatal results. People—professionals and residents—were at risk of first becoming unconscious and then dying of asphyxiation. In perhaps the lowest point of human history, HCN was also found in Zyklon B, the cyanide gas used during the Holocaust.

Like other inorganics, HCN use began to dwindle. "By the early 1940s," writes historian John Ceccatti, "field researchers had identified about ten agricultural pests and laboratory strains that were capable of withstanding exposures to insecticides that had previously been highly effective in killing these same species." Arsenic and cyanide became pointless chemical baths. But World War II helped introduce a juggernaut that brought attention to resistance. It was known as DDT, or the Third and most powerful Horseman of the Bugocalypse.

"The story of DDT," writes Dawn Day Biehler, "dovetails with that of postwar suburbanization, consumerism, and notions of modern homes." Housewives became generals in the War on Bugs. The compound, dichloro-diphenyl-trichloroethane, became famous for its comparatively *lower* toxic effects on mammals. The other side of that coin, as pointed out by Fumio Matsumura ("grand master of insect toxicology"), caused "environmental endocrine disruptions." Like the aforementioned pyrethroids, DDT thwarts an insect's sodium channels. And as with pyrethroids, insects evolved quickly to fight it. Unlike other insecticides, however, DDT brought international attention to resistance after it had lowered the volume of Earth's buzzy fiends for a short amount of time.

Disease-related fatalities dropped significantly after World War II because of DDT. Italy didn't experience one malaria-related death between 1949 and 1950. Any insect that came into contact with DDT was obliterated. By the mid-1950s, 100 million pounds had been produced, diluted into household products, powders, sprays, and infused wallpaper. Advertisements billed it as a "miracle powder." Section 8 homes were regularly treated with DDT. Over a 30-year period, the United States used over 675,000 tons domestically. The president of the American Association of Economic Entomologists estimated that 5 million lives were saved.

Then in 1962 an atomic awareness bomb dropped in the form of Rachel Carson's *Silent Spring*. "The insect enemy has been made stronger by our efforts," Carson said. "Even worse, we may have destroyed our very means of fighting." By 1948 scientists had already heard confirmations of a burgeoning worldwide resistance. But now DDT became the "cause célèbre that ignited the environmental movement," writes Will Allen. The ecological chain reaction caused reproductive issues for birds (such as thinner eggshells) and aquatic species. Until then, biological control of insects, first introduced by C. V. Riley, had gone by the wayside. But even after DDT's ban in 1972, undeveloped nations relied on the immediate action of pesticides as more elegant methods were not readily available. Again, the aftermath was dire. According to a World Health Organization report from 1989, pesticides were poisoning 1 million people a year worldwide—approximately 20,000 died. After the 2004 Stockholm Convention on Persistent Organic Pollutants, DDT was limited to vector control only.

The EPA finally regulated a good portion of such toxins with the Food Quality Protection Act of 1996. I leave the seat of the Fourth Horseman . . . unoccupied. Resistance, just like viruses, constantly evolves in our attempts to treat it. A certifiable weapon against insects will not come into existence in our lifetimes. However, what can lead the way, at least for portions of time, is a mash-up of biological control.

Many scientists believe that nature can combat nature and that synthetic chemicals only complicate matters. Even C. V. Riley, the "father of biological control," expressed doubts about Paris green. He and two other scientists drove out scale bugs from California citrus groves by introducing Australia's vedalia beetles in 1888. In the wake of this success, entomologists traveled the globe searching for beneficial insects. The enemy of my enemy is my friend and all that jazz. "Every insect has its predator which follows and destroys it," Carl Linnaeus remarked in 1752. "Such predatory insects should be caught and used for disinfecting crop-plants." He was right. Eighty-five such biological control agents were put on trial and implemented from 1920 to 1940. Up to the present day, there have been 1,200 such projects. At times, the "benefit-to-cost" ratio of biological control as opposed to pesticide treatment, according to two researchers, can be 200 to 1.

A similar citrus issue sprang up in Israel around 1938 with mealybugs, so the Palestine Farmers Federation shipped several parasitoid species from Japan, including encyrtid wasps. Within three years after cultivation and release they'd snuffed out mealybugs. Other such integrated pest management (IPM) techniques might include pheromone traps, crop rotation, or shorter production seasons (which saved Texas from pink bollworm in the 1930s). Russian entomologist Elie Metchnikoff introduced the use of fungal pathogen *Metarhizium anisopliae* to control grain beetles in 1878. Sprinkled over fields and active for 11 months, it ended up being more effective against the sugar-beet weevils than any other means. Metchnikoff later moved to Paris's Pasteur Institute to continue researching entomopathogens to control pests, such as green cabbageworms. Another disease, called Wipfelkrankheit,* used to control insects, killed swaths of nun moths.

Despite IPM's slow but valuable controls, synthetic organic insecticides boasted more immediate results. From 1964 to 1982, such chemical usage increased 170 percent in weight, according to a paper in *The Economic Journal*. A survey from the late 1980s found that 40 percent of farmers were dissuaded from using IPM simply because they were uncertain of its effectiveness. According to University of Arizona professor Bruce Tabashnik, incidences of insect resistance to pesticides went up by 61 percent from 2000 to 2010. (That's from 6,617 records of resistance to 10,661.) "There's a reason [insects] have in essence conquered the world," says Tabashnik. "In terms of their numbers, their diversity and their widespread geographic distribution . . . This is a never-ending race. This is a never-ending conflict."

The increasing attraction to IPM has turned farmers on to transgenic crops engineered with genes from natural bacteria. In 1911, *Bacillus thuringiensis* (*Bt*) was found to kill flour moth larvae. And in the past two decades, since *Bt* became commercially available, farmers have gravitated to crops engineered with it. Such GMOs comprise over 79 percent of corn and 84 percent of

---

* Try saying that five times fast.

cotton crops in the United States. But if mishandled, moth larvae can become resistant. This is where Bruce Tabashnik comes in. To annihilate seed-eating pink bollworms in Arizona—a century-long "scourge"*—Tabashnik helped devise a plan incorporating strategic placement of *Bt* crops. Here's the twist. Rather than leave it to *Bt* alone, the plan pushed the use of sterile insect technique (SIT)—a process developed in the 1950s—to wallop male moths with radiation.

"A combination of techniques can be much more powerful than just their sum," he says, talking like a military colonel. "They work together synergistically." Once ingested, the *Bt* bacteria "punch[es] holes" into the gut lining of caterpillars. In the rare case that a resistant moth finds another resistant adult, "then they're off to the races." But by combining *Bt* crops with SIT, the chance of that outcome is reduced significantly. Starting in 2006, collaborating with the USDA, cotton growers, and the University of Arizona, the team began releasing sterile bugs. (Problems arise if SIT is used alone. In one instance involving codling moths, reduction tapered off around the fifth year of releases.) To detect if the technique was working, traps were set up, similar to Oxitec's GM mosquito project. "Don't rely on one thing," Tabashnik says about our future pest management. "That's a bedrock principle." What it takes, he says, "is a combination of tactics."

Pink bollworm in Arizona will be declared officially eradicated once the USDA gives the word. Yet *Bt* is a limited promise. Tabashnik believes his group's success should last decades. But human fallibility and insect evolution are, again, inevitable.

---

*Bruce Tabashnik e-mailed me a mobile phone video of what an infestation looks like. Colleagues in Gujarat, India, used the same *Bt* crops but didn't have the resources to properly implement them. Within 10 years the worms had adapted. The 51-second clip opens on mounds of frost-like hills of harvested cotton. It pans to a noisy cotton gin with fresh bolls hailing down. Above that is a detritus pan and, inside, a squirming pool of rosé-colored worms.

Each type of *Bt* is a narrow-spectrum bioinsecticide. Some types of the bacterium kill only caterpillars, or only beetle larvae or mosquito larvae. Other dubious foes include aphids, mealy bugs, stink bugs, and other sucking insects.

Pam Marrone, an ex-employee of Monsanto in a pesticide industry full of heavy-duty chemicals, is at the forefront of organic insecticides. Marrone grew up in southern Connecticut on a mini-farm in the late 1960s, a child of the *Silent Spring* era. And one of her vivid memories is of the local gypsy moth caterpillars. "I would be standing in the woods," she says during a phone chat, "and literally insect poop would be raining down on my head. There were so many caterpillars munching." She recalls the midsummer forest looking as depleted as though winter had arrived early.

What happened next changed her life. Her dad bought a highly toxic insecticide called carbaryl. After applying it, the caterpillars came down. This was then followed by a downpour of ladybugs, lace wings, honeybees . . . "My mother had a fit," she remembers. Her father then returned home with *Bt,* and from then on Marrone pursued microbial insecticides "single-mindedly." Natural remedies were a part of her upbringing. Her grandparents used folkish methods imported from Italy and Poland. Her mother taught her at-home organic treatments like grinding hot peppers into a spray admixture. "Now it was becoming a science."

Organic insecticides are clearly a growing trend. Companies like Terminix seek alternatives to crude chemicals. EcoSMART introduced plant-based oil insecticide. DOW partnered with a pharmaceutical company to discover insect-killing microbes. Scientists now dedicate their lives to finding IPM solutions. At Monsanto, in 1983, Marrone was a "kid in a candy shop," exploring the potential of 100,000 different microbes and ancient pest control methods. Later, after screening 77,000 microbes, she'd start Marrone Bio Innovations. Though her company is small, they've sold 1.69 million gallons of microbial-based biopesticides as of

2015, treating 2.4 million acres and targeting the other pests that *Bt* can't hit. This is small potatoes compared to the $55 billion in chemical pesticides sold globally today. (Biopesticides account for $3 billion.) But it is a healthier broad-spectrum solution rooted, Marrone's research has shown, in old techniques.

Attempts at putting pests in a fatal chokehold have been around for millennia. A *Scientific American* article from 1848 talks about burning peat as a means of fumigating bedbugs: "Peat in burning gives out a singular odor, which is very disagreeable to some, but which banishes that pest to mankind from houses, the bedbug."

The fumigation practice of burning sulfur is first documented by Sumerians in 2500 BCE, and is again referenced in *The Odyssey*: "Bring sulfur, old nurse, that cleanses all pollution, and bring me fire, that I may purify the house with sulfur." Another fascinating example: Greeks and Romans using a boiling mixture of bitumen, sulfur, and amurca—sediment from unfiltered olive oil—to keep caterpillars from decimating their vineyards. Our agricultural elders, they used natural resources to prevent infestations.

I ask if Pam Marrone has ever heard of the *Geoponika*—a 20-book series that compiles Mediterranean agricultural practices possibly as far back as 1300 BCE. Not only has she heard of it, but her scientists experimented with some remedies 18 years ago. After some thought, I decided to give it a shot too.

The *Geoponika* describes the practice of tying captured bats to "tall trees," write researchers Allan Smith and Diane Secoy. Their screeches would deflect locusts from farms. Scattering powdered stag antlers on seeds was believed to repel worms. And an application of bear's blood, goat fat, or frog's blood to pruning knives supposedly deterred insects and larvae. Coating cabbages with amurca and ox urine was said to have promising results. Amurca also acted as a pseudo-mothball that protected clothes and helped prevent ants from invading when smeared on floors. Staking a horse's skull in a garden like a medieval scarecrow fazed caterpillars. That goes double for nude, menstruating women dancing barefoot through the garden with "unbound hair." An owl's heart hung over crops repelled snails and beetles. So did the tilapia fish dangling from trees. Some Greeks recommended burnt seashells for mosquito repellent, while certain African tribes daubed cattle urine on themselves with the same hopes. And in Japan one remedy for keeping away nighttime crawlies was to place soft seaweed around children's beds.

Magic and folklore are the origins of pest control. Truly, they are as comparatively absurd as practices used in the past 150 years. And as many farmers do today, prayer offerings* were made regularly. Still, the question at hand is: does any of it actually work?

Bitter apple spray was used by Greeks to fight off fleas. Goldmoss stonecrop flower was used circa 300 BCE to infuse seeds as a preventive measure. Arsenic was used in China as early as 200 BCE. A combo of milk-softened hellebore and arsenic worked against flies. Oil was sprayed on granaries to upend beetles. During the Renaissance, "tobacco infusions" were recommended as a means of controlling pear infestations. Tobacco of course contains nicotine, which is still used in insecticides dubbed neonics.

In the 1970s, Smith and Secoy examined the *Geoponika* as well as other ancient writings. Their 1975 paper "Forerunners of Pesticides in Classical Greece and Rome" yielded interesting comparisons. Hellebore contains insecticidal alkaloids. "Certain animal fats," Smith and Secoy write, could have prevented diseases transferred by knives. Olive oil could "mask the scent of the fruits." But the benefits of amurca were hard to see, although, they say,

---

* Aside from being the thunder god, Zeus was known as the Fly-Catcher. Apollo was also a go-to for fighting locusts. Seventh-century Muslims wrote and folded their prayers onto poles in the field to keep that swarming pest at bay. Same goes for one seventh-century German abbot utilizing the power of prayer and the staff of St. Columba.

"the principles of fumigation involving evil smelling compounds may have had some temporary effects in driving away insects." A more recent study of plant-based repellents examined the volatile substances in leaves. The same leaves continue to decorate rural villages today as they did millennia ago. Odor receptors in the southern house mosquito respond to a chemical found in spice plants called linalool and properties of eucalyptus oil just as they do DEET. Same with citronella. The only difference is, unlike DEET, the natural substances' odors evaporate faster. That rate, however, can be slowed today by advanced nano-sized emulsion techniques.

When it comes to the Greeks and Romans, though, questions of effectiveness remain. For this reason, I decided to replicate four control remedies. While I did not manage to acquire bear's blood, by doing a little scrounging around, I found enough crude ingredients for some hands-on experimentation with hopefully beneficial results. Here's the list of treatments I was able to compile with the help of University of Georgia entomologist Jason Schmidt:

- Olive oil mixed with crushed chrysanthemum flowers.
- A pretreatment of cabbage seeds with a sprinkle of ground deer antler.
- Venison blood–soaked seeds.
- A spray application of ox urine and amurca.

The first thing I do is visit a local tannery, where I find a box full of antlers and one very perplexed taxonomist. He replies with a puzzled "Oh" when I tell him about the experiment. I'm greeted by the same confused wariness at a butcher shop, where I buy the bloodiest meal possible: prepackaged venison for a stew. Living in an apartment with no garden space, I make a couple of calls to local farmers who respectfully decline to host these ancient methods. That's when I beg my friend and the illustrator of this book Michael Kennedy to go Greek with me. Amused to the point of giddiness, Michael agrees, and I drive up to his home in Kearney, Nebraska.

We run into a problem. Oxen, it turns out, are difficult to find. Ox urine is even rarer. (At first I thought olive oil sediment was hard, but with the help of UC Davis's Olive Center and Corto Olive producers, I have two cups ready to sow.) As a compromise, Michael and I spring for a cow. However, even then we have an issue: how do we source urine exactly? You can't simply sneak up on a sleeping cow and dip its hoof in warm water. So, we compromise again. We go for deer urine instead, considering the animal's decentralized usage in this experiment thus far. A quick drive to hunting retail mecca Cabela's, and we're stumped yet again. The customer service person, looking as if he'd come from the Australian outback, comes up and tells us it's not deer season. The least synthetic urine they have in stock is fox. But this is farm country, after all. "My best advice," he says, "is to just find a milk cow and start milking, because that's when they start whizzing." Sadly, this can't be accomplished with the beef cows on Michael's parents' nearby hobby farm. Also, I don't happen to be carrying a kiddie pool for the mini Niagara Falls that cows can gush. Desperate, I ask the Cabela's employee if he may have a cow pee hookup. "To be honest with ya, we don't have much demand for that," he kindly states, shaking his head. "I think if you go out and ask a farmer, they might just shoot you on the spot."

Favoring my life, I opt for fox urine.

Using a patch of rich Nebraskan dirt in the back of Michael's garage, we plant four rows of pregrown cabbage plugs. Two receive designated applications of olive oil and chrysanthemum powder, as well as fox urine and amurca. One is hit with an off-the-shelf, wide-spectrum carbamate insecticide, while another row is used as a control with no additional substances.

Four months later I phone Michael for the results. He has some unfortunate news. "They were trashed," he says grimly. Apparently most of our patch grew into full-sized cabbages. But these veggies fell victim to the nightlife of this college town. The back of Michael's home has an alleyway leading to Kearney's main street—a byway to bars, convenience stores, and fast food. Apparently these cabbages were showered with more than just a dose of fox urine. But before piss-drunk college boys felt it necessary to relieve themselves and stomp out the evidence, Michael tells me that the plants grew to be healthy and edible. The *Geoponika* worked! In fact, the only row of cabbages that failed was our control, the one with no insecticides, which started to "split open," he tells me. "Otherwise, I could eat any of them." After a thorough rinse, of course.

Two additional rows, grown from seeds pretreated with antler dust and deer blood, however, never sprung leaves as the cabbage seedlings died shortly after planting. Such is the trial and error Greeks certainly encountered on this road to pesticide development, its need obvious and beneficial when practiced wisely. As entomologist Robert Snetsinger wrote in his 1983 book *The Rat-catcher's Child*: "The development of a professional pest control industry was an important element in mankind's social development and is associated with an increasing concern for a better life on this earth and the ability to make improvements in human living conditions." And it will be endless.

The War on Bugs is a curious one, fraught with casualties. As with the benefits and repercussions of pesticides, there is a balance to be struck—one we could possibly reach as the wiser generations of the twenty-first century learn from past mistakes. We might want to holster that swatter every now and then, because insects most certainly do improve the conditions for human life. And, more important, death.

Six

First Responders

You can assign a dollar bill to almost anything—even nature. To help estimate how beneficial wild insects are (in terms of moolah), entomologists Mace Vaughan and John Losey arrived at this equation in their 2006 *Bioscience* paper:

$$V_{ni} = (NC_{ni} - CC_{ni}) \times Pi$$

They found this value by first subtracting the estimated pest damage to crops with current control methods by the "greater damage" from no control. The bug-men-cum-economists then multiplied the price of these "natural enemies" by the proportion of baddies snuffed out by insects. Plugging in a few more calculations revealed that pest-controlling insects saved $4.5 billion per year in the United States alone. That's icing to an annual $57 billion ecological cake. Their paper analyzed the ingredients and layers of such services, reaching that whopping price tag without factoring in man-driven insect commodities like silk or honey. On the top of that list was insects' contribution to wildlife nutrition: an estimated total of $49.96 billion.

"That was probably the most surprising value that we found," John Losey tells me over the phone, midlunch from his office at Cornell. "Looking into the fish, the birds, and the mammals and what they fed on was an eye-opener." They derived the number by taking into account the money generated by US recreational activities like hunting, fishing, and bird watching and then factoring in the base of the food chain: insects. But this is limited to one industry—a mere fraction of insect benefits. It took Losey and Vaughan a year of research to calculate the reported economic value. There's simply not enough data or an algorithm to quantify the other factors.*

"We can't live without insects in the long term," he says conclusively. "How long would we eke out a survival? I don't know."

Pollination aside, in a world without insects, a hanky couldn't possibly quell the global stink. According to Vaughan and Losey's paper "The Economic Value of Ecological Services Provided by Insects," dung burial prevents a whopping $380 million yearly loss to the cattle industry by busily disassembling feces and, as a result, recycling nitrogen. Mammals defecate about 40 percent of what they eat. Based on recent cattle head inventory from the National Agricultural Statistics Service and solid waste research from Losey and Vaughan, there's about 2 *trillion* pounds of poop a year. Therefore we must be grateful for these processing agents who eliminate 10 percent of our nation's refuse.

"The community of insects and other organisms in a dung pat is more complex than you might expect," reads a line from Gilbert Waldbauer's *What Good are Bugs?* Such tenants might include fly maggots and fungi-eating mites and springtails. But beetles take

---

*For example, insects can reduce the frequency of wildfires. Two researchers noticed that from 1970 to 1995 spruce budworm outbreaks in British Columbia "significantly decreased risk of forest fire" for seven-year periods. It is questionable, however, how much of a long-term benefit such outbreaks are for ecosystems.

a major slice of the (cow) pie within 15 minutes of it being laid by either burrowing chunks beneath the pat or taking it to go* in a round ball to later drink the "liquid 'soup'" with their soft mouth parts, as described in *Encyclopedia of Insects*. Losey and Vaughan found that about a third of US cattle dung can be recycled by these ultimate trash compactors.

So, what would the planet be like without these fecal heroes? I'm happy to say we'll never know . . . unless, of course, you were born in Australia 50 years ago.

In the jazzy bongo and bass opening of *Dung Down Under*—a 1972 government-produced documentary about how Australia solved its eighteenth-century bovine dung issue—a beetle is hind-limb-deep in excrement. In 1788, colonizing fleets introduced cattle to the landscape. However, native beetles hadn't evolved to reduce robust, moist droppings, as they were accustomed to doing with the dry, manageable pellets of marsupials. "One passenger was missing from the livestock consignment," the narrator says. "The dung beetle."

What happened over the next 200 years as farm industries grew was the result of untended cow pies. If a bull were to "exhaust" 10 pounds of fecal nitrogen in a month, one study explained, only one-fifth would get into the soil. Toss in dung beetles and the return is two-thirds. The fecal invasion also brought pests, and maggot-ridden pats decomposed slowly, sometimes taking years. The ever-increasing number of buffalo and bush flies and midges disturbed livestock, decreasing life spans by transmitting pathogens like *Salmonella*. And the hand motion of waving flies from your face famously became known as the "Aussie salute."

Enter Hungarian entomologist George Bornemissza. Escaping

---

*Clingy moths reside on three-toed sloths, waiting to slurp on their weekly dung droppings, and scarab beetles solicit around wallaby anuses.

the rise of his country's Communist Party, he found himself in Australia's Commonwealth Scientific and Industrial Research Organization (CSIRO) by 1955. He was the first to suggest the importation of dung beetles to create sustainable agriculture by cycling nutrients and reducing fly populations. Ten years later, the proposal received funding, and Bornemissza soon found himself in South Africa procuring suitable beetles, the most adept of which were found in Mozambique's Gorongosa National Park. As a test he placed pats in a controlled paddock and timed how long it took for various kinds of beetles to break them apart.* After picking the most effective species, scientists sent eggs to a quarantine lab in Australia where lab researchers crammed turds through a Play-Doh-like sausage stuffer and then hand-rolled them to house the beetle eggs from Africa. After the eggs were incubated, the scientists set the adult beetles to work in a lab setting using dung squeezed from bags. The time-lapse sequence of beetles melting away a cow pat (which reminds me of *The Evil Dead*) is worth a watch on its own.

Starting in 1967, 55 dung beetle species, 22 of which came from Africa, had been mass reared and introduced to farmers and pastures across northern Australia. Half of them proved successful, and according to a 2014 article in the *Sydney Morning Herald*, the Aussie salute has become a "dying custom." So dense was the *Onthophagus tarus* beetle population that males' horns receded over time as wrestling battles over females became unnecessary. By 1985, release of the tunnelers and ball rollers tapered off until just recently when two new species (*O. vacca* and *Bubas bubalus*) were set loose by CSIRO in hopes of doing some spring cleaning.

Bornemissza's legacy, as he said, was "to do something . . . no-

---

*Researchers at Lund University found dung balls to be "thermal refuge(s)" by using infrared thermography. Savanna surfaces can surpass temperatures of 140 degrees Fahrenheit, which makes lugging dung around a fatiguing sport. To cool their hot feet, beetles rest atop the shit "platform," decreasing their temperatures by 12 degrees within a matter of seconds. Because fresh dung balls are about 90 percent water, once the beetles are back on the sand, the dung tracks soothe them as well. The Swedish researchers went further by fitting silicone booties on beetles, which resulted in them taking fewer breaks.

body has attempted before," and he proved that simple solutions can make large strides in agricultural management, especially when it comes to recycling organic matter. For instance, clothes moths are adept at reducing fur and hair. Termites, also called "white ants," can recycle over 50 percent of dead plant material. Microorganisms within their guts enable them to process cellulose. Seemingly nothing slips past these tiny disposers of matter and waste, these consumers of the past. And that could potentially include man-made refuse.

In 2013, two Canadian researchers reported on the curious choice of building material made by alfalfa leafcutter bees in Toronto. Researchers observed bees constructing nests with polyethylene-based plastic bags, collected locally, and composing the brood-bearing cells and their doors. One row of cells showed the plastic had replaced 23 percent of the otherwise natural material derived from resins and leaves. The magnified images of the nests are visually rattling in that mutated monster kind of way.

Mealworms that grub on Styrofoam have also been proven to pitch in. Researchers in Beijing found that although polystyrene is "resistant to biodegration," *Tenebrio molitor* larvae not only had a hankering for it, but their gut bacteria could break down the long-chain molecules into monomers. "The mealworm gut can be considered an efficient bioreactor," pushing degraded polystyrene into the biomass, the authors wrote. We use and discard nearly 300 metric tons of synthetic plastic per year on a global level, according to the researchers. So there was a certain sigh of relief when, after two weeks, spectroscopes showed that nearly 50 percent "of the ingested Styrofoam carbon was converted into carbon dioxide, . . . reveal[ing] a new fate for plastic waste in the environment." (Though bits of micro-Styrofoam were left behind as dust.)

Mealworms are fast; just 500 can turn 30 percent of six Styrofoam grams into Swiss cheese in a period of 30 days. Like the adult Toronto bees, the larvae pupated and became darkling beetles. Don't expect dung beetles to pack and roll snowball-sized Styrofoam any time in the future. I simply promote the thought: if bees and worms can recycle nonorganic material, think how effective they are with the organic stuff.

The most important bug contribution, John Losey believes, is the incalculable one: the decomposition of dead plant and animal material. "Some people would say the bulk of the work of decomposition is with microbes . . . In a sense that's true." But maggots give it a head start by first stripping the rotting waste into mulch. What's that like? To get an idea, I traveled to Texas A&M University to a series of bungalows collectively known as the Forensic Laboratory for Investigative Entomological Sciences. Or for the short-winded, FLIES.

"The breadth of what they can eat is extremely wide," says Jonathan Cammack. We're sitting in a half-lit meeting room at FLIES. Outside in a nearby greenhouse, cages full of black soldier flies colonize piles of waste. By the time the larvae, aka maggots, crawl away to pupate, they're composed of approximately 40 percent fat. In fact, such fat stores can be seen on the "windowpane" backs of the adults, occasionally appearing as a radiant green. When their stomachs are empty, you can see clear through them, as I learn later pinching one of their wings. In terms of recycling food waste, the scientists at FLIES have been promoting this specific species for a good reason.

"They have the Midas touch," Jeff Tomberlin chimes in, joining Jonathan and me. He wears a burgundy sweater vest and a wide southern smile. Before grabbing a couple of Lone Star beers together later, I easily pictured him being a host on QVC. "One of the major issues with waste," he continues, "is that they produce noxious odors, greenhouse gases . . . By using the black soldier fly to convert wastes into protein and oil, it also reduces these odors."

Years have been spent proving how beneficial the flies' recycling techniques are. A paper coauthored by Tomberlin discusses how black soldier flies can cut manure waste in half all while reducing the *E. coli* it contains. Their maggots were especially voracious eaters when it came to kitchen waste. Salads, hamburgers, and other boxed foods from restaurants (extremely high in fat, calories, and proteins) were consumed 98.9 percent more than standard poultry feed. Though it did take more time, the maggots they yielded, which can fetch $330 per ton on the feed market, were "longer and heavier." For all their data, Cammack and Tomberlin's black soldier flies haven't taken flight just yet.

There is a hindrance: federal regulations prohibit insects as livestock feed, which is the larger part of how Tomberlin and Cammack want to utilize these maggots. However, other countries have embraced the technology. AgriProtein, a company based in Cape Town and backed by Bill and Melinda Gates, currently employs black soldier fly larvae for large-scale biorecycling of city waste and then as protein-rich feed before they pupate.

Part of Jeff Tomberlin's interest in this field originates from watching similar cycles as a child on his parents' Georgia farm. "I remember seeing a dead cow on the pasture," Jeff tells me. "And I was curious. So, I went up and hit it with a stick—and three dogs ran out of it!" He says it's one of those visions that stays with you. Jeff and Jonathan tell me a bit more about black soldier flies before Jeff turns to Jonathan.

"You should have David put his hands in a pan of maggots."

Me: "Sorry?"

Jeff grins. "Yeah."

Moments later Jonathan leads me past fridge-sized incubators housing various soldier fly experiments. Then we enter a dim storage room with the pans. I ask him, with all that intense munching going on, how hot can maggots get?

"When the ambient temperature was 90 degrees," he recalls, "the temperature of the maggot mass was 140 degrees."

I let that number—140—sink in for a second before submerging my hand into a pan of very alive dirt. On the surface, the maggots are tranquil and cool and sparse. Their posteriors look about three times larger than their heads, and they have two brownish spiracles, or holes, from which they breathe. But when I dip my hand into the bottom, the mound I scoop writhes with life. In my palm it feels disturbingly ticklish. And the very biological, metabolic warmth I'm getting—well, all I can say is I've stretched my yuck spectrum beyond donating blood to a jar of bedbugs. On the plus side, the dead skin cells on my hand are gobbled up.

"If the insects we study didn't exist," Jonathan tells me, "the world would be covered in dead things."

But this wriggling handful doesn't push my yuck-ometer needle into the red. There's a darker side to FLIES.

What I have not yet mentioned about doctors Cammack and Tomberlin is that they're both well versed in mediocriminal entomology. This means that should you be murdered or abandoned dead—let's just hope not—insects are your best friends. Forensic pathologists can derive time-of-death (TOD) estimates within 24 to 36 hours of finding a corpse. After 72 hours? You'll need to contact a forensic entomologist. And they are most definitely scarce. The total number of board-certified experts in the United States can barely fill a freight elevator.

On average, Jeff Tomberlin tackles about 10 murder cases a year. Tomberlin estimates TOD based on specimen vials, an autopsy transcript, crime scene photos and video, or weather station data. When I told him about my urge to examine the forensic marrow of that interplay of bugs that feed on our carrion—our putrefying flesh—he pointed me toward Michelle Sanford, who works a couple miles away from FLIES in Houston. Sanford is the world's first and only full-time forensic entomologist, working on an average of 60 cases a year.

Understandably, Jeff's lips are sealed on case details. But as we walk back to FLIES from a bar, he leaves me with an amazing story. One day Jeff was walking outside . . . "And this man came up, shook my hand and said, 'Thank you.'" His details of the run-in are vague for an obvious reason: Jeff had investigated the murder the man was accused of—strangling his mother in a meth-addled rage with an electrical cord.

People wonder where you go when you die. Heaven's pearly gates? Hell's fiery depths? Detroit?

But the right answer is simpler: you are masticated by maggots and play host to numerous insects. It's not as bad as it sounds. While the worms certainly do crawl in, they also crawl out. Bellies full, they pump that nutrition back into the world. When all of the teenage angst, collegiate self-discovery, 9-to-5 office time, soccer practices, retirement parties, and rollicking times in the old age home are done, the big payoff of life is to become a pulsating worm farm.

As the famed forensic entomologist Zakaria Erzinçlioglu* wrote in his memoir *Maggots, Murder, and Men*: "Viewed dispassionately, a dead human body is a magnificent and highly nutritious resource."

I decided that if I'm going to meet a woman who has "maggot collecting" in her full-time job description, then I'd like to see the source firsthand. When I asked Jeff Tomberlin why Texas is such a great zone to study decomposition, he said it's likely the warmth that lingers 10 months out of the year. That's why, just after winter, I'm standing in an outdoor body farm, staring as a throng of flies† emerge and return to the slit of a cadaver's dried mouth like bees to a hive.

Weather: 81 degrees Fahrenheit with a calm breeze.

---

* Whom police respectfully nicknamed "the Maggotologist."

† Before Raid's advent, the first civilizations were overrun with flies. Amassed garbage and dead bodies drew "legions" of them, writes entomologist Bernard Greenberg, ingraining their cultural presence early through depictions on ancient Mesopotamian cylindrical seals. A passage from the *Epic of Gilgamesh* notes, "The gods of strong-walled Uruk are changed into flies and buzz around the streets." Deceased Egyptians wore "burial beads" to slow their decomposition. Associations to the dead peaked

"She's still pretty fresh," says Lauren Meckel, my graduate student host. "But you can see the flies are attracted to her." I've driven 130 miles southeast of Texas A&M to a 26-acre plot of land in San Marcos called the Forensic Anthropology Research Facility (FARF). Here I'll witness insects working overtime. The female cadaver before us was placed only two days before. Now she's mannequin smooth, or "marbled," and less recognizable. Lauren tells me the speckled, "sawdust"-looking schmutz on her face are actually fly eggs.

Bugs erode organic matter beyond recognition, dissolving 60 percent of a cadaver's mass in one week, and can munch away at such speed as to heat a corpse to 122 degrees. A maggot's predictable nature can narrow a missing person's timeline well enough to correlate with other existing evidence to help identify them—unmarred face or not. The species give validity to the nursery rhyme "Who Killed Cock Robin?": "Who saw him die? I, said the fly, with my little eye, I saw him die."

Lauren rattles off a list of some of the crawlies she's seen at FARF: "We get blowflies, cheese skippers, ham beetles, gnats, flea beetles, spider mites, soldier flies, dermestid beetles, ants, big black beetles with hooks on them, butterflies who really like decomposition, millipedes, stink bugs . . ." All of which are important characters in forensic entomology.

Parisian army vet and morgue worker J. P. Mégnin expanded the methodology. His 1894 book *La faune des cadavres: Application l'entomologie a la medicine legale (The Wildlife of Corpses: The Application of Entomology to the Coroner's Office)* was a landmark in the growth of Canadian and US medicocriminal entomology. It documents nine stages of decomposition with the succession of the first insect responders to a dead body. Succession on a corpse, thoroughly unveiled in the mid-1960s by North Carolina gradu-

---

famously with pagan god Beelzebub, originally written in Philistine text as "Ba'al Zebub," which translates to "Lord of the Flies."

ate Jerry Payne, is the process where bug colonization attracts new organisms. His carrion work on baby pigs found that 422 arthropod species can usher you forward after death.

Corpse infestation can resemble a nightclub, with early guests, VIP rooms, and last call drinks. The first to show are fly families Calliphoridae and Sarcophagidae, which sniff out bodies from four miles or more. (In South Africa, in 1981, researcher L. E. O. Braack marked flies and found that they could track carrion from a distance of 40 miles.) Female blowflies lay eggs in a fresh body within 30 minutes of finding it. Days after bodily colonization has begun, flesh flies get a leg up by depositing maggots already developed from eggs. Given that the head's nostrils, mouth, and eyes are good places to lay eggs, matured maggots traditionally liquefy the brain first (or enter through bullet holes in homicide cases) and stew in the cranial vaults and nasal cavities as well as genitals (lending to the idiom of having bugs up one's butt).* Discarded or dead pupa shells (depending on outside temperature) clue in entomologists to the developmental stages and thus a more accurate estimate of TOD.

Over the next five months as the body dries, and depending on the season, necrophage predators run the show. Silphids, or burying beetles, detect the foul stench of skatole from our degrading, leaky intestines. Ideally, these carrion beetles will deposit larvae into a topsoil chamber near the body, regurgitating dehydrated flesh or cartilage into their offspring's mouth. Moths and beetles typically stay once the party is over and assist with additional cleanup.

The guests at the second body we observe have basically vacated. On one rib lies a lone cheese skipper maggot, a species

---

*Oodles better than having bugs up your keister while you're still alive. "Scaphism" is an ancient torture method in which victims are bound and fed milk and honey. The onslaught of diarrhea attracts the same carrion flies. Maggots amass in the anus—never mind the bee attacks. Soon after, gangrene spreads, as does your internal bug infestation. Yeah. Please leave the night-light on.

prone to jumping, so I distance myself.* "Watch out for the hair mat," Lauren warns.

I look at the ground at a deflated clump of hair. "Is that his toupee?"

It's his scalp. Maggots, as we've learned, work from the inside out. Removing fine layers of dermis, like this cadaver's hair, is part of that. Originally, I'd thought the bodies at FARF were wearing latex gloves before I noticed the busy bugs behind their translucent skin. Nerves kicked my perspiration into high gear. I can't tell if the smell is me or the dead bodies.

Lauren lifts the chicken-wire cage protecting a male cadaver from furry mammals. The bearded fellow has the gaping yawn of an opera singer, eroded further by decay with a gullet of maggots. Larvae of various instars (of which there are three larvae growth stages) crawl over one another. My host leans over the body. "Looks like there was some scavenging activity on the arms, so all that tissue is gone," she says. In its place the bicep has been carved out into a rugged bowl of churning rot. Maggots break the black liquid surface like excited dolphins. The skin has a tanned glaze you'd see on a golden pig at a Hawaiian luau. The visit ends after we approach the next body (as the gentle breeze dies down). Leaking from the left side of the cadaver is "a lot of purge," frothing with maggots. I stare at the chunky, enamel pool. It's going to be awhile before I can enjoy a creamy risotto again.

As we head out, Lauren mentions being relieved there were no cockroaches. *Wait, what?* I imagine these pools of rot versus the little guys from *Joe's Apartment*. "I hate even saying the word," she says.

So, I ask, "Is it because they're greasy?"

"*Eww!*—ugh," she cringes.

---

*Neurotically. I later learn the ground they can cover, at best, is 15 centimeters. So, no Olympic medals for the long jump any time soon.

I do a quick scan of the decaying backdrop surrounding us. "Was there an incident that happened in your childhood—"

"*Blech*," she says. "One fell into my bathtub. That was traumatic." Should Lauren pursue a career as a crime scene investigator, I hope she'll bring a brown paper bag to breathe into, as a cockroach— though typically not collected for forensic entomologists—could be telling in other parts of the investigation, like disproving alibis. For instance, in one case from the mid-1980s, a grasshopper with a missing leg was unorthodoxly collected and marked as evidence. When said broken leg was found in the jeans cuff of the suspect, though, it was enough to show he was at the murder scene.

But proving insects' importance in criminal investigations took convincing. When forensic entomologist Lee Goff became involved with the study in 1983, bugs—these dwellers in our body's "temporary microhabitat"—were discounted in investigations. "Insects were not then regarded as a significant source of information by most medical examiners . . . crime scene investigators or lawyers," he writes in *A Fly for the Prosecution*. Public interest in the field is attributed to Bernard Greenberg, an outlier since the 1950s. He refers to bugs as "winged bloodhounds." And the small faction of forensic entomologists who understood their use, like Goff, called themselves "The Dirty Dozen." Case after case, insects demonstrated their cunning as detectives, missing only a down-turned fedora and a .38 Special. Since the '80s, Goff has worked on over 300 cases, even consulting with TV producers on *CSI*. But bugs' dramatic role in helping sniff out murderers started centuries before.

A death scene investigator training manual written by Chinese judicial administrator Sung Tz'u in 1235 CE documents a murder where flies essentially fingered the suspect. A farmer had been decapitated, so locals were lined up to present their sickles to authorities. When flies gravitated toward the murderer's sickle, he 'fessed up. It's hard to believe; so in an effort to prove the story's legitimacy, the scenario was recently re-created by a student at Sam

Houston State University using blood cleaned from a knife. According to one professor, when the tampered blade was set out among normal ones, the flies landed on it within "seconds."

Modern forensic entomology helped solve a murder for the first time in 1850, writes Gail Anderson. During house renovations, a mummified baby's body was uncovered behind a chimney mantelpiece in a Paris residence occupied by four families in three years. All of them were considered suspects. It took the deduction of medical doctor and naturalist Marcel Bergeret to peg a rough TOD estimate. So he performed an autopsy and found the empty puparia of flesh fly larvae in several cavities of the full-term baby, which were laid, he concluded, shortly after its death in 1848. Normally several dozen species would occupy a fresh body. His discovery absolved three families who lived there and convicted the previous owners.

But the discipline was practically sidelined until a Chicago double murder occurred in 1976. "Father of forensic entomology" Bernard Greenberg's testimony on blowflies found at the crime scene narrowed the TOD window and helped put away the accused men. Today, "headline homicide cases," writes Greenberg, typically have two expert entomologists who testify for the prosecution and defense, delivering "minutely scrutinized" hypotheses derived from their evidence.

Bug forensics helped get Casey Anthony off the hook. Two professionals created a true standstill with the arguments they made, which can sometimes promote scientific advancement. In 2011, NAFEA member* Timothy Huntington testified for the defense and veteran entomologist Neal Haskell testified for the prosecution. The question was whether Casey had asphyxiated her two-year-old daughter Caylee in the trunk of her car or if the girl had simply drowned in a pool. Both men agreed on the trunk's putrid smell, but clashed on its origins. Haskell found that the larval stage

---

*North American Forensic Entomology Association, that is.

of the discovered maggots matched theories that a body had been stashed for three to five days. Huntington, however, believed the sparse organic materials in the trunk's garbage bag attracted them, and that a body—based on his past experiments on decomposing pigs in car trunks—would've attracted thousands of insects. "Based on the findings," testified Huntington, "there's no reason to believe there was ever a body in the trunk."

Interestingly enough, Houston's Michelle Sanford is working side by side with a DNA analyst on a project that, if available during Anthony's trial, might have changed the result.

Whether we're bug infested or not, there's a stop most of us will make on the way to the cemetery, and it's where this twenty-first-century DNA analysis is being tested. The morgue at Harris County's Institute of Forensic Sciences bears an odd resemblance to an auto repair shop. There's a funny smell of burnt electrical circuits and chemicals and the sound of an occasional buzz saw (cutting through bone). And if you look through the smudged windows in the hallway, you'll see a series of corpses being disemboweled. Michelle Sanford does a different type of diagnosis.

Until 2013, chances are the insects colonizing the dead might've been tossed in the trash or "smashed" by the autopsy doctors. But then Sanford, who was taught by Texas A&M's Jeff Tomberlin, joined and slowly began to prove her worth. Insects have answers too. "In the beginning it was like, 'Please collect something. *Please,*'" she says about the first responders to a scene. Now the forensic specialists are taking plenty of bugs into consideration. (One time Tomberlin comically had a roly-poly sent along as evidence instead of the bevy of maggots.) After a couple of training sessions, people in the medical examiner's office have become more aware of what insects to collect. In a sense, Sanford is a prototype for a new trend in the field since she works full time on cases. As of 2016, she's worked on over 300 deaths—both homicides and natural causes. As the number of these specialists continues to grow, they may become a standard in investigations. "I've always had an interest in solving problems with insects," she says, rarely speaking above a loud whisper. "Forensic entomology, on a smaller scale, is very much like that."

Read an entomologist's extensive case notes, and you'll find a meticulous fiend at work. Notes may cross-reference National Weather Service meteorological data with indoor temperatures; measure maggot samples within millimeter decimals; and re-create environmental conditions with raw meat (usually pork dressed in clothes given pigs' hairless resemblance to us).

I pore over some of Sanford's consultation cases. The first is about a woman in a "5th wheel trailer" who died of morbid obesity. Judging from the instar size of bronze bottle fly larvae, Sanford estimated 72 hours had passed since the body's discovery and the woman's unfortunate demise with lines such as "additional maggots were observed on the couch cushions where purge fluids had collected under the body." Another file tells of a septuagenarian attacked by honeybees while doing yard work. Sanford mentions the removal of 20 stingers from the poor guy. At the end of each report there are notes on raising pet maggots from the strangest origins to help identify the species and find out how old they were: sheep blowflies "reared from [the] couch [and] ashtray," and flesh flies "reared from [the] nose."

Later, Sanford shows photos from the interior of a deceased hoarder's house. "It's full of insect attractants," she recalls. "They didn't want us walking on the second floor because they think we'd fall through." Not only can insects creep indoors, but Sanford's seen them in offices on the seventh floor. (Actually, a 2015 research paper from Malaysia recorded such high-rise colonization, finding evidence that insects reached one body 11 stories up.) The opposite of this is true with coffin flies, which dig up to a foot beneath soil to reach dead things. Next are images of the bald head belonging to the old man attacked by bees. The next slide shows a murder victim: a bullet wound springing forth blood-glistened maggots.

None of this made it to Sanford's Career Day presentation at Carter Lomax Middle School in Pasadena, Texas. Instead, she showed photos of a turkey she used for studying decomposition at Texas A&M. A couple of students loved the presentation enough to consider joining the field. An elementary school student once wrote her a thank-you note: "My favorite part was when you told us that when a person dies, flies can lay their eggs inside." Below her comment was a crayon drawing of orange maggots encircling a fly. Jeff Tomberlin took an entomology course while working at a funeral home. Robert Hall told the *Los Angeles Times* in 1989: "It requires someone who's just a little bit odd to begin with."

It helps in making odd requests, say, to research toxicological extracts* taken from maggots. In what Lee Goff calls entomotoxicology, drugs affect the developmental stages of maggots ingesting cocaine- or heroin-dosed bodies. To test how high the worms got and if it sped pupation, Goff acquired lethal amounts of cocaine in 1987 with the help of a Hawaiian coroner in order to drug rabbits. The scientists kept tranquilizer darts nearby should anxiety overwhelm the coked-out critters. However, the varmints ODed before the tranquilizer guns were unholstered. Goff's maggot research then showed that larvae grew much more rapidly when aided by cocaine. Still, although they obtained pupation earlier, the coke maggots reached adulthood concurrently with the control group.

Molecular DNA extraction in larvae or windowpane specks (that's fly poop and vomit) also provide potential evidence. Checking the gut content, say, of the phorid fly maggots in 2008's Casey Anthony case might've supplied the DNA evidence to separate

---

*No blood and urine required! In the late 1980s, maggots collected from a 67-day-old cadaver—"brain and testicles not found"—were washed and homogenized and analyzed through a liquid chromatographic procedure for a postmortem drug test. Whether or not the maggots experienced a drug-addled daze is unknown.

food from garbage versus human flesh and convict Caylee's murder suspects.

Working with Sanford on this novel technique at the Harris County morgue is DNA analyst Chaquettea Felton. "We've been playing with different techniques," Felton tells me, "like using an insulin needle to inject into the head of the maggot and take out whatever we can." She and Sanford have also tried crushing the maggots for extraction or dissecting their guts. "Cutting the crap out [of them] works pretty well," Felton clarifies. "But using the needle to suck it out actually works better. So once we get the insides out," she says, giggling, "we do a pretreatment." Afterward she purifies the extracted sample and amplifies the DNA through a polymerase chain reaction to make copies. By performing what's called a capillary electrophoresis, Felton can separate a human profile from the sample.

She holds up a chart measuring relative fluorescence units—an analysis of the processed gut content that better resembles stock market trends. "The peaks you see here are where you get amplification at these different regions." Those peaks signify the DNA profile of the dead. "And so we were able to take out the maggot insides and actually get the profile from the human it was eating off of, which is pretty cool."

Sanford has been taking maggot samples from people who died of natural causes in an effort to build consistency for this new approach in forensic science. So far, the best samples have been from corpses found indoors with plump flesh fly maggots. I'm certain by the time of this book's publication, Sanford and Felton will have validated the process, making it a handy tool in homicide investigations.

And for the record, like our friend Lauren at FARF, Michelle Sanford dislikes cockroaches. As she put it: "That's where I draw the line."

Though the idea of an insect VIP party in your body might make you squeamish, I hope their nutrient cycling can impart some solace. Still, you don't have to be a corpse to get friendly with maggots. They have been used in medicine for centuries. Recall the barbed stingers Sanford plucked out of our unlucky Texan. Stings from the order Hymenoptera (e.g., wasps, hornets, etc.) account for the majority of insect-related deaths in America. Ironically enough, that very venom may improve on today's drugs and save us in ways never before imagined.

Seven

You Just Squashed the Cure for Cancer

If you've ever been stung, you probably know it's not as fun and cutesy as the preschool song "Baby Bumblebee," in which children pretend to clap the assailant. Had they really clapped a bumblebee, they would enter a world of pain via the bee's muscled, stinging apparatus. Nerves in the stinger would continually inject venom into their hands, which spells out big-time ouchy. Daredevil entomologist Justin O. Schmidt indexed the intensity of bee, wasp, and ant stings (observed naturally or induced on himself) from 21 species in 1984, ranking the pain from 0 to 4, from "no pain" to "traumatically painful." Schmidt clearly described the sensation caused by the "mechanics" but was fuzzy on the "chemical nature" of the pain and damage—a mystery, he mentions in his recent memoir, *The Sting of the Wild*, that chemists still have not solved. How much pain can humans endure? For a reference point, honeybees ranked at a 2. Bullet ant venom, however, caused a sensation like "walking over flaming charcoal with a 3-inch nail embedded in your heel." To date, Schmidt has been bitten by 150 species (which has luckily saved others from revisiting the experiment). But when it comes to venom qualities and other buggy secretions, there are daring individuals who seek promise rather than peril. Getting stung is something worth singing about.

Entomotherapies, the medicinal uses of insects, have existed for millennia. Some Westerners, though, remain skeptical of the process—with good reason. The long history of entomotherapies seesaws between early medicine and straight up superstition.* In India, carpenter ant jaws have been used to suture gashes since 3000 BCE. The Ebers Papyrus—an Egyptian medical guide from 1500 BCE—lists a recipe for psychological afflictions: "Take a big scarab, cut off its head and wings, boil it, put it in oil and apply as an ointment to the affected person's body. Then cook the head and wings in snake's fat and give it to the patient to drink." Dioscorides, a Greek physician circa the first century CE, wrote that crushed bedbugs crammed up the urethra combated bedwetting. Twelfth-century Germans, according to a Benedictine abbess, injected ants still very much *alive* into people suffering from fatigue. Elizabethans mixed rabbit urine with powdered earwigs for a concoction poured into the ear canals of deaf people. Red ants' acidic secretions† assist wounded Thai as an antiseptic. And a case of chicken pox in China? Try a cattle tick prophylactic. Actually, to date, China has concocted over 1,700 different drugs from arthropods. And nearly one-third of Hong Kong residents drink royal jelly bee extract, which clinical trials found good for depression.

But the true medical secrets of these insects—excruciating

---

*Unless your name is Louis Armstrong. To prevent his voice from getting too gravelly at a young age, Satchmo ate cockroach soup. Sound hokey? Consider the ground cockroach paste used as a "miracle drug" in Chinese hospitals. "They can cure a number of ailments," Professor Liu Yusheng told a British news reporter. "And they work much faster than other medicine."

† English naturalist John Ray remarks on this "Acid Juyce" in a January 13, 1670, letter to the publication *Philosophical Transactions*. He notes how remarkable the extract is: "I doubt not but this liquor may be of singular use in Medicine. Mr. Fisher hath assured me, that himself hath made trial thereof in some diseases with very good successes."

Hymenoptera venoms included—are only now being tapped. Drug discovery in nature from the 1960s up through the '90s shifted focus toward plants. "Insects," Georgia entomologist Aaron Dossey tells me, "fell into the cracks." His research, entitled "Insects and Their Chemical Weaponry," proceeds to lay out hypothetical game-changers. A similar paper, entitled "Bugs as Drugs," written a couple of years earlier by Miami doctor E. Paul Cherniack, notes: "Although medical practitioners in more economically robust countries may prefer conventional treatments, it may be more a result of squeamishness rather than science."

South Americans had known this for some time. When arthritis became too much to bear, they'd stick their inflamed joints into the "tree of the devil" where ferocious Hymenoptera fire ants (*Pseudomyrmex triplarinus*) nest. Natives voluntarily received venomous bites to send their rheumatoid arthritis into remission. Roy Altman of the University of Miami milked that extract for a double-blind study in 1984. Sixty percent of the venom-treated patients experienced a significant reduction in their swollen joint index. Four years ago, researchers from Brazil's University of São Paulo State analyzed fire ant venom and identified 46 proteins—a promising potential for new drugs. We now run mass spectrometry over bugs to analyze their chemical makeup. In fact, there's currently a US patent on Hymenoptera venom for pharmaceutical production as treatment or prevention of nerve-damaging and autoimmune diseases, which include arthritis and multiple sclerosis.

Honeybees have a proven track record in medicine.* Korean acupuncturists using bee venom–coated needles relieved more pain than traditional therapies thanks to proteins that reduce the *ouch* factor. Ancient Egyptians treated wounds with anti-inflammatory

---

*In fact, mead—a delicious, honey-fermented alcohol—used as an ancient potion, carries some interesting etymological ties. "[Mead] is the basis for the word 'medicine,'" writes entomologist James Hogue, "in recognition of its purported healing properties."

insect derivatives like propolis—a honeycomb glue made from plant resin. Folk healers fought canker sores with propolis as well, which has been proven effective recently in randomized tests. Today, *in vitro* studies show an inventory of implicit boons in propolis's polyphenol, acidic chemicals. Tuberculosis antibiotic? Check. Leukemia suppression? Double check. Leukemia has also been treated with an age-old Asian medicine: blister beetles—natural producers of cantharidin.* This chemical was found to suppress growing bladder, colon, and oral cancer cell lines, E. Paul Cherniack reports.

Impressive as *in vitro* studies may be, modern, marketable progress has been made in the field. Some American hospitals use inventions related to folk healing practices. The company Medihoney creates bandage dressings utilizing the osmotic properties found in honey that, like propolis, are anti-inflammatory and moisturizing for burn wounds. Such gauze has been found to reduce heal time by half versus the standard treatment of silver sulfadiazine. Additionally, there is the company called, appetizingly, Medical Maggots. Their containers of sterile maggots (*Phaenicia sericata*) are placed—this time on the living—in patients' post-op wounds and plugged with gauze and tape. Larval therapy debrides necrotic tissues and has also improved non-healing venous ulcers, as well as abscesses and gangrene. A study done on 86 patients showed a "66- to 100-percent reduction of wound size." (Big thanks goes to the antibacterial secretions from the maggots' guts.) Those claims explain the 5,000 lab-grown, myiasitic maggots delivered to US hospitals every week in the 1990s. About 20 are needed to treat a square-inch wound. A 2007 study promoting their use estimated 50,000 bottles worth

---

*The main property in the sexual arousal tincture known as Spanish fly. The infamous aphrodisiac—derived from blister beetles—has a centuries-long history and is used to this day. One case in 1996 involved a group of Philadelphians admitted into the emergency room hours after consuming Spanish fly–spiked drinks. The compound caused urinary tract hemorrhaging and bleeding, as opposed to the sexytime they'd hoped for.

of "medical-grade maggots" were delivered to hospital patients in 20 countries.

Mayans made use of the medical benefits of maggots, as did soldiers in Napoleon's army wounded on the battlefield. Those attended by maggots had a better chance of leaving with their limbs intact. This isn't bad, considering the numerous beatings that bugs delivered to Napoleon in his pursuits of conquest. "Although these insects were troublesome," wrote one of Napoleon's army surgeons, "they expedited the healing of wounds by shortening the work of nature, and causing the sloughs to fall off." And beginning with the US Civil War and later World War I, myiasis was used to treat open infections thanks to the work of William Baer. The French emperor was also aided by another bug, though not in war. Leeches* literally saved his butt.

Napoleon had a classic case of hemorrhoids. (Conquering nations is stressful, okay?) But a shtickle of Preparation H wasn't yet available. Leeches, which leave a mark said to resemble the "Mercedes-Benz emblem," writes hematologist Amiram Eldor, have saliva with antihemostatic agents that "delay clot formation." You get where this is going. It's harder to imagine the predicament Napoleon's physician faced. *Now if you would, dear dictator, kindly remove your breeches and spread your cheeks . . . You may feel a slight nip.* This sort of therapy has proven a success for arterial repair and is growing in use in areas such as plastic surgery where improved blood flow helps engorged organs.

Venom, which in medicine functions as a yin and yang, is also making its way into operating rooms. Take deathstalker scorpions. Responsible for hundreds of deaths a year, the world's most poisonous arachnid may also be used as a new standard in brain tumor surgery.

---

*Leeches—praised for their anticoagulant saliva—first appear in Egyptian hieroglyphs dating to 1567 BCE, and earned the name *Hirudo medicinalis* in the Roman Empire, and were used cosmetically—almost to the point of extinction—by Victorian women, who stored them in lavish fine-grained marble jars. Back then the hype was known as "leechmania."

At the PopTech convention in 2013, Seattle brain cancer researcher Jim Olson projected an image of a cancerous canine brain tumor illuminated in fluorescent wonder midsurgery, shocking the crowd. Deriving the peptide chlorotoxin from Israeli deathstalker scorpions, Olson reengineered the cancer-binding protein to act as a distinguishable "flashlight" for neurosurgeons, who normally accidentally extract healthy brain tissue while trying to ensure they've removed all traces of cancer. Precautionary gray matter removal impairs patients' neurologic function—an unfortunate necessity in this game of millimeters. But Olson's "Tumor Paint"—scorpion peptides capable of penetrating the *im*penetrable blood-brain barrier—guides surgeons "100,000 times more" accurately than MRI scans. The peptides also target the lymph channels in which cancer cells travel undetected during surgery. Gliomas and astrocytic tumors become surgical "road maps" in organs, ablaze in vibrant green hues.

Human clinical trials began in December 2013 in Australia. US trials began in late 2014, one site being Los Angeles's Cedars-Sinai Medical Center. Various studies have also found that the scorpion's chlorotoxin was a potential visual guide for other body parts, including breast, liver, kidney, prostate, and lung tissues. But Tumor Paint, because it was a bug derivative, seemed far-fetched. Initial grants submitted to national health institutions were turned down for being "highly speculative," Olson told the PopTech crowd. But families who had faith in the pediatric oncologist donated $5 million to the research. More compounds are emerging. Flublok, a flu vaccine with a three-week production rate (as opposed to the typical six months) that was recently approved by the FDA, uses viral DNA from fall armyworm moth ovaries.

Olson's separate venture, Project Violet, aims to design drugs from chemicals found in plants and animals, repurposing proteins as Olson has done with Tumor Paint. Currently his researchers, in addition to University of Queensland scientists, are examining the cancer-killing proteins in funnel web spiders, one of the world's deadliest arachnids. Insects may also end our antibacterial plight.

Using the same deathstalker scorpion derivative, virologists in China modified the peptide to kill *E. coli* and MRSA bacteria—the latter causes the ever-contagious staph infections in hospitals. "Drugs are encoded in their DNA," Olson said. "And they've had millions of years to evolve."

Biochemists have other crawlies under their microscopes. "Earthworms," E. Paul Cherniack writes in *Alternative Medicine Review*, "have a rudimentary immune system and contain antimicrobial and antineoplastic substances." One, called eisenin, "destroyed" cancerous tumor cells in humans. Prialt, derived from a snail venom peptide known as ziconotide, has become a painkiller substitute for morphine, especially when morphine tolerance becomes an issue. The drawback, however, is that unlike chlorotoxin, ziconotide can't pass through the blood-brain barrier; it therefore has to be injected directly into the spinal column. A study from the Washington University School of Medicine in St. Louis shows that melittin in bee venom may offer hope for couples with one HIV-positive partner who'd like to have children. Melittin was shown *in vitro* to break through "the protective envelope that surrounds [the] virus" and destroy its basic structure, leaving cells in sperm and vaginal tissue lining unharmed.

Meanwhile, the number of superbugs, i.e., multi-drug-resistant bacteria, has swelled over the past decades due to our habitual overuse of antibiotics. Twenty-three thousand people die annually in the United States from infections for which multiple antibiotics exist. However, a group of researchers from the University of Nottingham found substances in the brains of locusts and cockroaches that kill superbugs. Because cockroaches are coprophages, i.e., poop eaters, they evolved hardy antibacterial agents.

This shifting, twenty-first-century view of insects is not only affecting medical science, but robotics as well. If we look back at the history of technology, the list of bio-inspired devices goes on and on.

"Technical innovation," writes Professor Gerhard Scholtz of the Humboldt University of Berlin, "is sometimes the product of the observation of nature." He attributes the invention of the wheel to dung beetles: "This rolling activity is one of the most amazing actions in the animal world and forms a combination of various techniques and a distinct art of craftsmanship and engineering . . . The combination of rotation around an axis, making use of the low friction resistance of circular and smooth surfaces to transport a heavy load, shows the closest degree of similarity to a wheel that I can think of."

Insects are proven catalysts for spurring innovation. As that lovely old-timer Aristotle once mused: "If one way be better than another, that you may be sure is Nature's way."

Recall the Spanish song "La Cucaracha," about an unlucky cockroach? To help me cross the scientific intersection of insect and technology, I've enlisted a tan and blotchy *Blaberus discoidalis* to undergo transformation into an RC car. Now, this might sound like a painful, RoboCop moment, but unlike the five-legged roach limping around in the Mexican ditty, the *B. discoidalis* will return to normal once again (minus the addition of some headgear). The cyborg conversion kit I've purchased, called RoboRoach, comes from Ann Arbor, Michigan, startup Backyard Brains.

The rewired neural impulses work like this: an electrode, a red circuit board, and a battery are glued onto a 10-centimeter cockroach. Silver ground wires inserted into the roach's antennae stimulate its sensory neurons. When it receives a "spike" through your smartphone's remote control interface (via the company's app), the roach gets the impression it has come into contact with an object and turns left or right. These are actions (or pulse widths) you can fine-tune. In a research paper entitled "Line Following Terrestrial Insect Biobots," scientists Tahmid Latif and Alper Bozkurt aptly describe this type of locomotive control system as "similar to steering a horse with bridle and reins." Ultimately, the goal of Backyard Brains' RoboRoach is to educate kids. The electrical brain stimulators guiding the roach are similar to the ones engineers use in cochlear implants and to stimulate neurons for people with Parkinson's disease.

I call in my special order for *B. discoidalis* to a reptile pet shop and go pick it up. Given my love for the legend, I name him Bill "Fucking" Murray. (He even shares the actor's sense of irony and steely eyes.) Out on my apartment balcony, I saw the condensed brick of dirt substrate in half, dump it in a salad bowl, and add water. Voilà! Instant habitat. Plop a leaf of romaine lettuce into a terrarium, and Billy boy is living happy.

Now to address the screaming little PETA protestor inside you. Consider Billy's brief biography, with 15 or so siblings produced a couple times a year and a likely encounter with the bottom of a shoe if seen in public. Put together, steering Bill "Fucking" Murray with my iPhone via the gigantic, Bluetooth-enabled circuitboard he lugs around—clumsily staggering like a drunken backpacker—gives him worth. And it's educational. P. B. Cornwell says in his 1968 book *The Cockroach*: "It is a reasonable assumption that more cockroaches have been dissected on the laboratory bench than any other insect." We'll soon see why. Also, rest assured that after his procedure and test run, Billy will live out the rest of his cockroach life (approximately 20 months) with me.*

Beethoven's Symphony No. 3 plays through my computer speakers. Inappropriately, it's on the second movement—a funeral march. I've dunked Bill "Fucking" Murray in ice water, inducing hibernation. Surgery tools—toothpicks, Silly Putty, and tweezers—are laid across the table. After his legs stop kicking, I remove him from my beer mug and use a Q-tip to swab water from his head's shield, called a pronotum. With sandpaper I remove a waxy layer from his exoskeleton, upon which I then Super-Glue the electrode header.

---

*Andy Warhol, during a 1980 dinner party conversation with William Burroughs, is quoted as saying: "I used to come home and I used to be so glad to find a little roach there to talk to . . . They're great. I couldn't step on them." My sentiments exactly, Mr. Warhol.

Resubmerging him in water, I wait before inserting the "ground" wire into a hole poked in his thorax and clipping his long antennae down to a quarter inch. Seeing as I'm wearing blue surgeon's gloves and listening to classical music, I can't help but hear Peter Seller's Dr. Strangelove narrate the procedure. *I vill now take ze antennae and clip zem halfway in order to insert ze electrode wires.*

My hands are shaking. The wire is only 0.003 inches in diameter, so it feels like inserting a hair into another hair. The difficulty in accomplishing this and successfully Super-Gluing it in place makes what happens next emotionally trying. In my haste to prep for the operation, I've forgotten hot glue gun sticks, which are meant to keep the tangle of wires in place. Billy, in an innate reaction to obscurity, tears out the wires from his antennae while I sleep. Soldering is the only way to repair it, and I'm not entirely ready to do that to him.

So, I decide to get a bunk mate for him and retry the experiment with a cockroach named Archy.* I replicate the surgery days later—this time with hot glue—and let Archy rest overnight. But even with secured wires, the roach, perhaps with the help of Billy, has pulled them out. Tenacious creatures indeed. Deciding not to put them through any more RoboRoach surgeries, I vow to augment my RC bug another time.

Others have succeeded where Archy, Bill "Fucking" Murray, and myself have failed. This similar repurposing of cockroaches is

---

*Taken from *Evening Sun* columnist Don Marquis's fictional character from 1916 in *Archy and Mehitabel*. The satiric column asserts that one morning he "discovered a gigantic cockroach jumping about upon the [typewriter] keys" with verses expressing the woes of the everyday bug:

> swattin and swattin and swattin
> tis little else you hear
> and we'll soon be dead and forgotten
> with the cost of living so dear

under way at North Carolina State University, where backpacked roaches will become responders to disaster zones. The idea is to use them to swarm and explore areas where GPS can't penetrate, mapping out the detail of, say, the rubble in a collapsed building— tunnels, gaps, voids—using radio waves emitted from their sensors. Randomly scavenging dark places is nothing new to the roach. The reason some engineers have studied terrestrial insects as potential for biohybrids for 25 years is their deft access to the inaccessible and adaptability to various topographies.

"In engineering," says Professor Alper Bozkurt, the electronics whiz behind the project, "the first thing you do is look at nature and find out how [problems] are solved." Tiny robots that mimic insects are ideally what cyborg roaches aim to be one day. But, as Bozkurt points out, their current physical limitations prevent meeting that micro scale and, even more so, micro power supply. With bionomics, however, we've seen remarkable examples of the marriage between biology and modern technology, he tells me. Cochlear implants. Cardiac pacemakers. Bionic limbs. That's why insect bionics, harebrained as the idea may sound, is so attractive. You're dealing with a nimble, resilient creature. "You have a biological organism," he extols, "that can survive a lot of challenges, environmental predators . . . So we're trying to overwrite their instinct."

Directing the movement of cockroaches (or preventing them from dislodging wires from their heads) is as hard as it sounds. Bozkurt's colleagues at NCSU are figuring out a way to prevent their neurological assimilation—a point where their deep-rooted instincts kick in, thus returning them back to free-moving cockroaches. Over time, their response to electrode stimulation gradually loses efficiency. Bozkurt says it fizzles out anywhere from 10 minutes to 10 weeks later. So the methodology requires some tweaking. One includes liquid metal electrodes, which he compares to the T2000 in *Terminator 2*. "We open up a small hole in the antenna, and inject a liquid metal inside. Sometimes if the tissue dries, you don't have a good connection."

Yet Bozkurt's team has managed to assimilate an entire group of roaches. In a demonstration, they steered the micro herd on a circular platform. Each was controlled individually, but to the audience it appeared to be a swarm. While a real-life deployment of their roaches may be three or four years down the line, Bozkurt's inbox is flooded by the eager requests of various search-and-rescue operations and military organizations. By then, Moore's law should scale down the tech, creating systems lighter in weight and giving his team of roaches more maneuverability.

But Frankenstein-like biohybrids are merely a step toward biomimicry.

Cockroaches as search-and-rescue teams are nothing new to Robert Full. For over 25 years, Full has studied animal locomotion for the sole purpose of building highly adaptable robots for various topographies: climbing slick walls, navigating surface debris without having to slow down, paddling on top of water as if taking a leisurely stroll. The Berkeley-based Poly-PEDAL Lab named their ever-evolving, bug-like construction RiSE (Robot in Scansorial Environments). It integrates aspects found on the ever-elastic cockroach, like bendy spines, toes, claws, as well as the dry adhesives found in a gecko's foot pads used to climb and support itself without magnets. Full has run cockroaches through simulations that mimic explosions and earthquakes, as well as over vibrating gymnastic high bars, further exploiting their moxie. Scientists, including Erich von Holst, have made great headway on studying insect gaits since the 1930s. They have discovered their legs are "governed by independent control systems," writes biologist Holk Cruse, "each step with its own rhythm." This grants extraordinary freedom of movement, which is why so many inspired engineers are applying such mobility to robotics.

Insect elasticity inspired Danish zoologist Torkel Weis-Fogh to study the wing inertia of locusts in the 1960s and '70s. Although he initially analyzed their aerodynamics in 1951, his later experiments employed nascent stroboscopic light and high-speed cameras. Fogh captured the "clap-and-fling effect" of bug flight, which is crucial to understanding the low-pressure pockets their motions make for thrust, lift, and drag. Such motions cause wake capture—an aerodynamic force built off of the wing's previous rotational stroke. And fruit flies, a recent study showed, can bank turns at 5,000 degrees per second. A *PLoS* paper coauthored by zoologist Simon Walker describes how their "dramatic flight maneuvers" are attributed to the "13 pairs of steering muscles" discovered in blowflies using microtomography (read: a fancy X-ray, 3-D-generating computer camera). "Deformations of the [fly's thorax] wall," the paper concludes, "are not only responsible for transmitting forces from the power muscles to the wings, but are also important in accommodating qualitative changes in the modes of oscillation of the wing articulations." These deformations might influence future flight design.

It's no wonder the Defense Advanced Research Projects Agency (DARPA) is funding the development of micro air vehicles (MAVs). The MAVs have come far from the dive-bombing, insect spaceships of the arcade classic *Galaga*. DARPA intends for these centimeter-long, agile MAVs to one day recognize faces, hover, detect biochemicals,* and swarm and kill—ahem—subdue enemy targets completely unmanned, claims one Air Force promo. "They'll help ensure success on the battlefields in the future," says the stark voiceover to heighten the drama. "Unobtrusive, pervasive, lethal . . . MAV, enhancing the capabilities of the future war fighter." Alper Bozkurt's bionic approach to MAV was less intimidating; it involved 72 MHz AM transmitters implanted into tobacco hornworm moths. According to his 2009 paper, super-regenerative receivers were surgically inserted into

---

*The US Army explored toxin- and bomb-sniffing skills with arthropods from 1963 through 2006. A kinder substitute to dogs. In smaller laboratories, other entomologists, like Jeff Tomberlin (chapter 6), have successfully conditioned wasps to detect chemicals. The wasps are placed in an observation container with a webcam called the Wasp Hound that delivers live feed from the inside as a technician watches the insect's response to recognized smells.

pupae. Later, the adult emerged with an "adopted implant" in its thorax and the moths were tied to helium balloons. The crazy thing is that it worked. The balloon helped with the payload of additional power sources, cameras, and sensors—all while the moth was controlled remotely. It's questionable how efficient this might be in terms of mass production. But it's an impressive step toward a long-term goal: autonomous flight synchronized by a unified system.

If the early tech adopter in you wanted a MAV, engineers at Air Force–supported TechJect offer one based on dragonflies. The desire to mimic dragonfly flight rests on one obvious feature: two sets of wings. Because of this "improved aerodynamic efficiency," dragonflies can hover, fly backward, and travel in air at low speeds, all while reducing the energy of normal insect flight by 22 percent, says a 2008 study. TechJect has produced new MAV iterations every year since 2012. Their goal was to build a one-ounce, pocket-sized, four-winged drone capable of things like aerial photography and security monitoring. The crowdfunded project was hit with a couple of complaints from supporters after failing to deliver in a timely manner. It ended in 2015, though similar ventures often spring up.

Success and a giant leap toward the ideal vision of MAV bio-mimicry can be found in a Harvard lab—the Promethean birthplace of the RoboBee. The 2013 *Scientific American* article "Flight of the RoboBees" puts the achievement succinctly: "Their tiny bodies can fly for hours, maintain stability during wind gusts, seek out flowers and avoid predators," write the researchers. "Try that with a nickel-sized robot." It's mainly composed of two flapping wings and a flat carbon-fiber airframe cut from ultraviolet lasers and then folded into shape like a "children's pop-up book," describes Harvard professor Robert Wood. It weighs 80 milligrams, less than an actual honeybee. It is the result of nearly two decades of progress, largely inspired by the massive bee die-offs occurring still today. The hope is that thousands of them will mimic an actual colony and even go on rescue missions—possibly in tandem with cockroach ground patrol.

RoboBees copy, in a way, the same "wing-thorax mechanism" as the aforementioned blowflies. But the researchers encountered a couple of snags, including brittle actuators and, like Alper's roach backpacks, voltage issues. This is why the RoboBee currently requires wires to tether it to an external power source. The upside to this is that the dynamics of the actual "brain" of the robotic insect can be crafted, including prototyping different camera systems to one day reach the same optic flow visual recognition bees use. A more recent advancement using electrostatic adhesion has enabled the robotic insects to perch on leaves, wood, steel, brick, glass, and other surfaces. This downtime can perhaps allot a battery recharge in between flights, just as insects do. By the time you read this, RoboBees will undoubtedly have learned new tricks.

Engineered biomimicry, bioinspiration, and bioreplication of insects is for the most part a recent development. That is surprising, considering their intuitive design. Remember the ventilation ingenuity of African termite mounds from chapter 2? The subterranean South African nest has a closed-off chimney capable of releasing the humidity from breeding (which can range from 90 to 99 percent). These thermoregulation constructs are a true marvel, with numerous "porous" holes on their sides permitting cool breezes to enter. As wind wraps around the structure, a vacuum emerges, sucking out the warm air and helping maintain the nest's temperature at 87 degrees, a temperature optimal for the termites' cultivated fungus. Architect Mick Pearce copied that design for Zimbabwe's Eastgate Shopping Centre. Over a five-year period, the self-cooling structure built in 1996 saved $3.5 million in electricity costs.

In his book *The Shark's Paintbrush*, Jay Harman goes into some of the bug-inspired devices that have developed rapidly over the past 10 years. San Diego–based Qualcomm is attempting to utilize the crystalline structures found on vibrant butterfly wings to create remarkably energy-efficient TV sets. Panelite designed insulating glass with honeycomb-mimicking hexagonal structures, which can be found at JFK Airport, to diffuse light and reflect solar heat and thus lower air-conditioning costs. Companies like Bolt Threads and Spiber Technologies are racing to perfect biomaterials that will re-create the durable and strong web material of spiders.

Consulting groups like Biomimicry 3.8 and PAX Scientific, of which Harman is CEO, are thrusting those bio-inspired inventions into the everyday, taking cues from nature for industrial designs. Bugs, once ignored, are now innovating improvement in a number of sectors. To name a small fraction: one group of researchers built 180 microlenses into one new camera lens for undistorted, 180-degree images inspired by insects' hemispheric eyes. And it turns out that the exoskeleton scales on beetles scatter visible light wavelengths. Mimicking such fibers can give us whiter teeth.

But while sparkly smiles are nice, I'm tipping my hat to mosquitoes for my favorite invention to emerge from this biomimetic trend. You never feel them suck your blood. Perhaps you feel their spindly legs on your skin, but serrated parts in the proboscis minimize contact with nerves, making for painless blood withdrawals. Kansai University mechanical engineer Seiji Aoyagi draws inspiration from mosquitoes. The jagged surface of his silicon-etched hypodermic needle mimics the mosquito's "stinger." It has two outer shanks to penetrate the skin, while the 0.1-millimeter, vibrating needle slides smoothly to take blood undetected. Aoyagi says the needle is still "brittle" so human clinical trials haven't happened just yet. But for the 20 percent of Americans with needle phobia—yo!—this offers the potential for enormous relief. And it's one of many advances that may hit the marketplace soon.

Technology and medicine are literally emerging from the cracks. How could biomimicry and insect derivatives not? Insects' ingenious designs have evolved over 400-million-plus years. So of course they are lending their six or eight legs in helping us to create better TV sets, surveillance drones, and antibiotics and are guiding the hands of surgeons. With bugs, discoveries and monetary benefits abound. The show-and-prove moment is no longer necessary. Need proof still? Just ask the plethora of businesspeople who've made billions off these buggy assets.

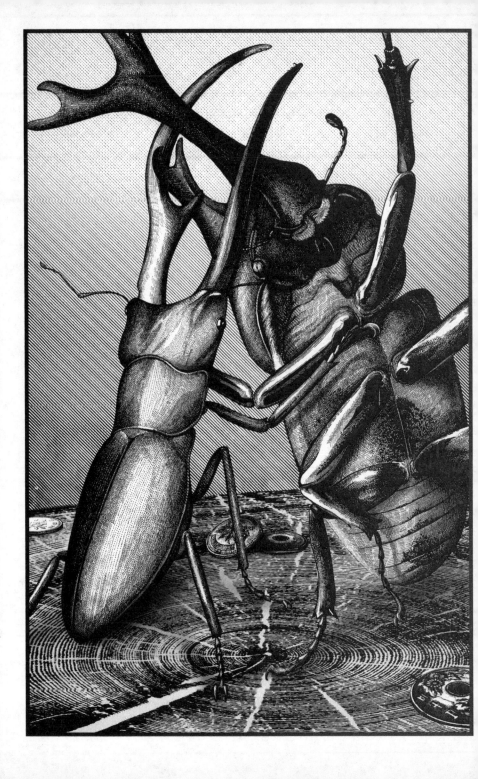

Eight

Executives of Big Bug Biz

"**M**othra oh Mothra."

These words are the first line of prayer sung by teeny twin demigods in the eponymic 1961 film *Mothra*, named after a fictional Japanese monster, aka *kaiju*, capable of blasting hurricane winds and spewing destructive silk. *Sayōnara*, Tokyo. But history has shown that the praiseworthy, winged goddess—who later became Godzilla's ally (why not?)—also holds significant roots in Japan. How deep are those roots? Esteemed entomologist Nan-Yao Su points our attention to a group of people in a village around 644 CE in the eastern Shizuoka Prefecture who worshiped and performed rituals of song and dance . . . to a caterpillar.

The ancient book *Nihon Shoki* referred to this caterpillar as "Tokoyo no Mushi." The phrase means "Insect of the Everlasting World." It was this green-colored, citrus tree–crawling caterpillar that villager Ōfube no Ō advised his fellow neighbors to praise. Local witches proclaimed that "those who worship . . . will, if poor, become rich, and if old, will become young again." This makes me wonder just what other animals or objects were ousted from the deity list before the villagers arrived at a caterpillar. Throughout history societies that attribute mystical progressive technology to others have been known as cargo cults. They consist of groups of people responding to immigrants from "more advanced culture[s]," writes Nan-Yao Su. The catalyst in the case of Tokoyo no Mushi bore a symbiotic relationship with citrus trees brought from China. Praying to the plant's caterpillar, imbued with this "magical potency to produce wealth and prestige," would in turn advance the primitive Japanese cult.

"Insects play a big, big role in symbolizing a certain kind of emotion and aspect of Japanese life," Su tells me. Why they chose the caterpillar as their deity mascot* is tied to the Shintō religion's "animistic belief" that spirits live within objects found in nature. This characteristic, Su points out, is seen in Hayao Miyazaki movies like *Princess Mononoke* in which wise wolves and deer evoke spiritual connections. Other animistic traditions still practiced today include a festival called *mushi oi matsuri* to ward off crop-damaging insects. In a procession akin to a funeral march, a line of farmers wave torches. Afterward, a memorial service for the dead bugs known as *mushi kuyoo* is held as lantern balloons ascend into the night. Whether or not people attended any observance for Ōfube no Ō is debatable. My own guess is the highly revered caterpillar failed to deliver on its promises, hence the villagers' "cast[ing] away their possessions." Because of this discord billowing throughout the cult, Ōfube was executed by a rather ticked-off Hata clan chief.

But Japan's prestige and wealth would arrive 1,200 years later through its dedication to another worm.

"Mothra!" Su roars, munching on pizza. "When *Mothra* came

---

*Not unlike the buggy idols honored in US sports arenas. We have the New Orleans Hornets, San Antonio Scorpions, Columbia Fireflies, Savannah Sand Gnats, Salt Lake Bees, Fort Wayne Mad Ants, and, my personal favorite, the Santa Cruz Banana Slugs.

out, I was like, 'Oh, yes! My insect.'" Its gargantuan dimensions and scope is appropriately sized considering the impact silkworms had on the country in the late 1800s. "Silk single-handedly saved Japan from being colonized," exclaims Su. "The reason is because [the Japanese]"—like a moth—"were able to transform themselves very quickly. Because of the silk! Silk saved Japan." Like the poor followers of Tokoyo no Mushi, nineteenth-century Japan was comparatively left in the dark ages. The Shogun ministers realized as much toward the end of the Edo era in the 1860s as US Commodore Matthew Perry forced open their ports to trade. With the Meiji Restoration came the national government's Charter Oath, which declared that "knowledge shall be sought throughout the world." After this declaration, importation began and machinery, industrial weapons, and foreigners introduced new technology. "The minute they realized how advanced European technology was," says Su, "*BOOM!* They change their opinion overnight."

Simultaneously, Europe's silk-producing caterpillars (*Bombyx mori*) were hit with a contagious microsporidian disease called pebrine. The disease toppled the industry in France and Italy. Although the economic threat was later subdued by findings from Louis Pasteur's new microbiology studies, at that moment, the silk industry needed saving and fast.

We tend to overlook the economic positives bugs provide: Successful industries like shipping butterflies and exotic bugs to botanic gardens and zoos across North America, farming crickets in giant warehouses, or airdropping pests into foreign agricultures as biological warfare.* The aphrodisiacal qualities found in some insects, like the fungus yartsa gunbu, which is grown from the

---

* Such was the proposed mission of the malumbia caterpillar. In 1990, the Bush administration entertained a USDA plan to sabotage Peruvian and Bolivian coca fields by infesting them with this relative of the gypsy moth. Although much of the cocaine would pass through the caterpillar, it would retain about "53 nanograms per gram of body weight," writes entomologist May Berenbaum. Perhaps justifiable enough to dry, dice, and snort the critter. Out of national pride and wariness of the agricultural repercussions, the plan never left the airplane's bay door.

heads of Himalayan caterpillars, can also spark successful industries. (The fungus's popularity—traced back to fifteenth-century
Tibetan scrolls—and price tag, about $50,000 per pound, have
led to the occasional theft and murder.) The caretaking of insects
has a long history, evidenced in cockfight-style cricket matches in
seventh-century China. Still, that doesn't touch the potential
goldmine raw, insect-derived materials exhibit.

In the twentieth century, East Asia became our foremost
source* for silk. In terms of the global market, Japan became the
top exporter by the 1930s, comprising 80 percent of the share. "In
1873 China exported three times as much raw silk as Japan," writes
economic historian Debin Ma. But this would change. A Frenchman by the name of Paul Brunat proposed a plan to the Meiji government in 1870 to build what would help, in part, advance Japan
into the technology giant it is today: the Tomioka Silk Mill. Soon
there'd be an increase in the number of local sericulture farmers
rearing silkworms. "From 1926 to 1933," writes Ma, "the share of
cocoons sold through this system grew from 12.5 percent to
40.1 percent." As China's conservative government with its Self-
Strengthening Movement—the polar opposite of Japan's new philosophical approach—stifled the growth of its silk industry, Japan
entered the modern world through its devotion to insects.

Rows of churchlike pews face a small video projection on the museum floor. We're in the Tomioka Silk Mill in Japan's Gunma Prefecture. I'm joined by my Airbnb host Ayomi Anabuki-Browning
and her teenage daughter Azusa, the three of us watching an im-
personator of nineteenth-century French engineer Paul Brunat.
Replete with a Western bow tie and white David Bowie suit, the

---

*And according to folklore dating back to 2640 BCE, Chinese empress Si Ling-chi was
sitting under a mulberry tree when a cocoon dropped into her teacup and unspooled
from its hardened state into thread. Evidence actually dates our cultivation of silkworms
to before 4700 BCE.

lean Frenchman looks like a slender, alternate-reality version of Colonel Sanders. Of course I doubt whether Brunat or Sanders spoke Japanese as well as this dubbed version of the mill director.

Chipper music plays as the camera pans across a CGI re-creation of the factory's 300 steam-powered mechanical reels, which were fully operational by 1872. The film cuts to a large brass bowl in a factory. Boiling alkaline water softens the silk of the cocoons—a composition of proteins known as fibroin and sericin, generated by the silk glands. Both are products of nitrogen from ingested mulberry leaves—a common tree here in the Gunma Prefecture. Fibroin is the main component of silk, so the binding gum sericin is dissolved in the hot water as the female workforce pull thin filaments out of the bowl and catch the yarn onto a reeling machine. Some cocoons contain nearly 3,000 feet of string. When faced with daunting quotas, Ayomi tells me, women would hide their children beneath the table to eat any excess, leftover threads.

"They don't show the dark side," Ayomi says, disappointed. "Where's the dark side?"

Azusa looks bored to tears.

Lightly Battered Colonel Paul Brunat Sanders walks the outside of the cocoon warehouse talking about—from my audience assessment—nothing particularly interesting as hundreds of stoic-faced women are shown. Then, suddenly, Brunat bids adieu by sparkling into radiant gold light. Yep. Somewhere a clock must've struck midnight because he turns into a moose-turd-sized cocoon right near the gate entrance where I just bought my ticket! Unable to personally plop Brunat's cocoon home into boiling water, Ayomi, Azusa, and I proceed through the facilities.

The three of us enter the second story of the east warehouse. This 300-foot-long timber construction with a 50-foot-high ceiling is where dried pods were kept. Numbers on the banisters denote the storage sectors. "It's like the same as Costco," Ayomi says. A museum guide tells us that the mill closed in 1987. At that time, the sericulture industry was making waves in Japan. Cold chambers slowed the growth of eggs to keep production going through the seasons. Wind caves functioned as large-scale egg banks like Gunma's Arafune Cold Storage, which maintains near-freezing temperatures during the summer. Lab studies produced transgenic moths for disease resistance and bigger cocoons. One way to yield more silk is to lengthen the last larval instar by playing with caterpillars' hormones. In their fifth and final instar (larval growth stage) before pupating, "they'll eat mulberry leaves like crazy," says Nan-Yao Su. "So the idea there is to expose them to juvenile hormone and delay their [pupation] by one or two days. It forces them to eat more." This pursuit of perfecting one's craftsmanship is very Japanese. Su, a Taiwanese man who spent his formative years in Japan, says the phrase for this is *shokunin katagi*.

By 1965, automatic reeling machines replaced the workforce. However, the 400-ton iron water tank that fed water to the old steam machines remains. It sits four feet off the ground on concrete blocks. We meet a more talkative man patrolling it. I ask if he can tell us anything that isn't included on the tour. You know . . . insider silk mill scuttlebutt. He ponders some. "Hmmm . . ." And then he raises a white gloved finger and spills the beans.

"Ah," interpreter Ayomi starts. "Local people, when they had kids, would come and use this as a playground underneath the tank." The kind gentleman sees the dismayed expression on my face—I can't help it. So, he mentions another thing. "So the French guy," Ayomi continues, "the founder of this factory, he used to drink red wine, right? For Japanese, they'd never seen red wine before. So," they thought, " 'it's red drink. What is he drinking?' " She feigns the nervous voice of a silk worker. "They think he's drinking *blood*. Human blood. That's why Japanese girls were scared of him. 'The Beast drinking the blood!' " They also believed he cooked with human fat. Ayomi giggles. "Their imagination goes too far."

Colonel Cannibal Sanders kept a large wine cellar. Actually, he lived here with fellow business managers before moving out by 1875. The houses and dormitories would welcome foreign dignitaries, including the Empress Dowager of China. For symbolizing such a cornerstone in Japan's economy, the Tomioka Silk Mill was listed as a World Heritage Site in 2014.

We stop in the village for *taiyaki*, fish-shaped cakes filled with sweet azuki beans, and then pack into Ayomi's car. As we drive through the narrow village streets of Annaka, I think back on the yellow cocoon pods I saw in the mill's gift shop. They were not, as I assumed, dyed. Rather, certain wild silk moths produce yellow pigmentation when consuming carotenoid-rich plants. Green cocoons exist thanks to metabolites known as flavonoids. Japan's National Institute of Agrobiological Sciences (NIAS) is where such mutant potential is being exploited. Although annual silk production has steadily declined over the past 20 years, sericulture's shokunin katagi idealists have not faltered in their pursuit for perfection. This is why we now have transgenic silk worms fused with the hardy dragline proteins from spiders, and neon-colored silks engineered for mass production. The red and orange fluorescent silks produced by NIAS are "10 percent weaker," the researchers said, than commercially bred varieties. But that should be enough to withstand Electric Daisy Carnival partying.

I look this up at Ayomi's house, the vast windows of my bedroom opening up to Gunma's mountains. It's the same natural light that filled the silk mill. In fact, I learn that silkworms were once reared on the second floor of this converted farmhouse. Ayomi and her British husband Simon have gone to great lengths to modernize this Akima home originally built in 1938. The lush hills of Annaka, like many parts of Gunma, were once heavily populated with mulberry trees; the landscape has now been replaced with plum orchards. After dinner, Simon and I take their golden retriever Hanna for a walk through the orchard. The crisp evening air is rejuvenating, and Simon tells me how patches of mulberry trees remain to mark property borderlines.

❧

The potential boon from various insect-related industries relies on our ability to abandon our aversion. Monarch butterfly migrations helped spur the ecotourism movement as guided tours take vacationers along their trail. There was a time when for 10 francs you could watch fleas drag Monopoly-sized tokens around a tiny circus ring. Long before being a superb polishing agent, beeswax was used as a currency in Rome. Iron gall ink, used from the time of the writing of the Dead Sea Scrolls through Bach's concertos and into the twentieth century, comes from oak tree galls created by a secretion from egg-laying cynipid wasps. And until 2012, a certain type of red bug juice gave your Starbucks frappuccinos happier hues.

"There is a tiny scale insect," writes scientist Stuart Fleming, "whose sole love in life is to dwell on the fleshy green joints of a prickly pear cactus . . . whose bodies owe their distinctive scarlet coloration to the carminic acid in their blood and muscular fibers." For a time, the top commodities of the New World were gold and silver and this liquid red dye. Before it became commonplace, its rare and regal color was highly sought after, upping its value. A kilogram of this dye, derived from female cochineal (*Dactylopius coccus*) by squeezing 100,000 of the pruney, sesame seed–sized tropical insects, can sell for $40. Together, Peru and Chile produce 220 tons of extract a year. But it's at Kentucky's D.D. Williamson coloring house where chemistry wizards tamper with the by-product's pH levels to generate a palette of hues, labeling it on food and cosmetic products as E120, carmine, or Natural Red 4. Manufacturers of the carminic acid saw market prices increase eightfold from 2009 to 2010. And the industry remains steadfast despite Change.org's successful petition against Starbucks' use.

The by-product's popularity began with the Mixtecs of southern Mexico. Ancient nobles wore clothes dyed in the "color of the gods" to denote their social status. Traditional farmers would harvest clusters of cochineal from prickly pear cacti using woven, cylindrical nests. The red dye's introduction to the New World and expansion of our sartorial minds is thanks to colonial Spaniards. The country's ministry besieged Spanish conquistador Hernán Cortés in 1523 to diversify and expand the dye's production. It gave Spain a monopoly for 250 years. But in 1777, a Frenchman snuck these cactus pads back to Europe to begin production elsewhere. The reign of cochineal's regal red extract as a commodity second to silver began to waver. The Spanish government moved cochineal farming to the Canary Islands, writes Fleming, exporting as much as 6 million pounds by 1875. Less costly, if less radiant, artificial dyes overtook the market toward the end of the nineteenth century.

The most ubiquitous of all extracts is the one that coats your jelly beans, nails, apples, pharmaceuticals, and coffee tables with a waxy sheen. You know it as shellac. The waxy resin secretions of the *Kerria lacca*, cultivated in places like India and Thailand, appear as tumorous blisters on tree branches. Its use as a dye goes back to 250 CE. In 1596, European writer J. H. van Linschoeten ventured on a scientific trip to India and described lac's usage: "They spread the Lac upon the whole peece of woode which presently, with the heat of the turning [melteth the waxe] so that it entreth into the crestes and cleaveth unto it, about the thicknesse of a man's naile ... The woode ... shineth like glasse ... In this sort they cover all kinde of householde stuffe in India, as Bedsteddes, Chaires, stooles, etc."

He *continueths* for a while about its "beautie and brightness." Lac harvest can employ millions of people, a fact that has remained largely unchanged over the past century.

Lac workers called *kharadi* still pick and melt the red clumps. They then flatten the resin through a wringer before breaking it into flakes. A liaising salesman known as a *baipari* goes village to village acquiring parcels of lac. *Baiparis* then take it to shopkeepers, or *arhatiyas*, who sell it on commission. After refining methods improved, its popularity increased. Varnishes contain about 25 percent or more shellac. Before vinyl came around, some phonographic records were made of insect resin. *Shellac* author Edward Hicks notes how the resin "stiffened" bowler hats, polished smoking pipes, styled hair, coated playing cards, and layered electrical wires for better insulation. Until the 1950s, its replication dumbfounded lab scientists. By 1998, India produced around 85 percent of the world's lac, exporting 30,000 tons per year worth $4.8 million.

Silk, shellac, and red dye. All have helped give rise to global powers, but these products are now in great decline. Before getting hopped up on future insect services, entrepreneurs should first possess an amicable relationship with bugs. Examples of such healthy relationships can be found today on the fringes of culture.

We once again turn to the East for enlightenment.

Gunma Prefecture sits smack dab in the middle of Japan. A landlocked, humid mass of Eden. Nestled within it is a 118-acre bit of habitat dedicated to the pleasure of viewing bugs—1,400 species of them, to be precise. This popular zoo of mini-monsters known as Gunma Insect World, built in 2005, also serves as a centerpiece for the widespread Japanese cultural phenomenon of bug-loving.

The local Tōbu Kiryū train achingly pulls into the modest Akagi Station. The zoo is a quick cab drive away. Waiting on the opposite platform is a character performer—a brown, bipedal pony that seemingly rolled out from the Hello Kitty wastebasket. This cherubic equine is Gunma-chan—the prefecture's mascot. Beetle horns sit atop its bloated head like a kabuto samurai helmet. He waves to the three kids starfishing their hands against the train windowpane. Once the door opens, the kids bolt over to hug this plushy character, their tiny backpacks bouncing jovially. I follow next for my hug and tourist photo, naturally.

In Japan, insect-loving kids are called *konchu shonen*. Their bug, or *mushi*, obsession is an itch Animal Planet documentaries can't scratch. "When I was a child attending elementary school in Tokyo," writes Akito Kawahara in an *American Entomologist* article, "my father bought this book." It was a "butterfly treasure map," he reflects, one they'd follow every Sunday, and Kawahara was shocked to find groups of butterfly aficionados "with their 30-foot extension poles lined up against a forest just after dawn," collecting butterflies. In the late 1990s, such mushi enthusiasts' purchases averaged "tens of millions of dollars" a year. In 2004, import trades to support the habit of konchu shonen were estimated at $100 million. Ushering that support was SEGA's arcade/trading card game *Mushi-King*—a role-playing, fighter video game using actual beetle species as models for the playable characters. I gave this a try while visiting the geek paradise known as Akihabara. At the basement level of the six-story arcade tower known as Club SEGA, three teens gather around a new edition of *Mushi-King*. I sat at the machine next to them and began smashing blue, red, and yellow buttons going fisticuffs with some beetles. At the end of each fight, a new character card with attributes I selected was printed off.

I love Japan. Where else can you collect 7-Eleven insect toys? The 1970s TV series *Kamen Rider* featured a motorcycle-cruising grasshopper humanoid. Today, department stores and markets sell live rhinoceros beetles as pets. Fifteen years ago, some vending machines even spat out live beetles. Bugs are the G.I. Joes of Japanese pop culture. Part of this infatuation is owed to the idyllic language of French entomologist Jean-Henri Fabre. His 1879 book *Souvenirs Entomologiques* (Entomologial Memories), explains Kawahara, was translated and published in multiple editions here. This shared obsession for beetles traces back to a temple in Nara

Prefecture and its seventh-century Tamamushi Shrine with 9,083 emerald and red-striped hardened wings.* Ditto other insects. One interpretation of the country's eleventh-century name, *Akitsushima*, is taken to mean "Dragonfly Island." All the manga and anime I've read reference or depict insects in various ways. There are the more direct ones like *Bug Boy*,† *Read or Die* with villainous insects threatened to be unleashed, and then *Insectival Crime Investigator Fabre*—named after the French luminary—as well as *Professor Osamushi*.

When I get to the Gunma Insect World ticket kiosk, though, the zoo is nearly empty. This is Disneyland on a rainy day during school finals. That's because it's mid-March. The Japanese colloquialism for how the weather's behaving is "three days cold, four days warm." The grass in the park is brown and the pathways are sparse with baseball-capped konchu shonen. But clearly I'm not as despairing as the silk tour guide visibly bummed by the off-season lull.

A middle-aged employee named Sada smokes a cigarette in front of a 140-year-old silkworm farmhouse. He's supervising about 20 kids running around the thatch-roofed piece of history that reminds me of Ayomi's home. Children play outside with Hula-Hoops and stilts. The farmhouse, filled with museum-worthy silk-reeling machines and beautiful tatami rooms, is entirely empty. I show patrolling guide Sada a modicum of enthusiasm for the silk artifacts in the house, and he breaks out into a dance.

"Wait," Sada says, "one minute."

---

* Iridescent elytra is sewn into dresses or set into jewelry. Unsettling too (certainly a fashion faux pas) are the very alive and bedazzled beetles some tourists smuggle into the US. Worn as gold-chained brooches called "Makech," the jewels stem from a lovelorn Mayan fairy tale.

† Of which there was a shirt I threw my money at. I also mailed one to our insect-pinning artist in New York, Lorenzo Forcella. This was his e-mail reaction to the *Bug Boy* shirt: "I love the larva eating the dead baby. It's funny because there was some chick in my neighborhood who didn't know my name so she just called me 'Bug Boy.'"

We place our shoes before the stairs that lead to the tatami room. He swings open a large closet, rummages around, his dark green wind jacket swishing, and returns with a card bobbin of twine. "Watch," Sada implores, his English almost as limited as my Japanese. He squats beside a low table with an authentic, Edo-era manually operated apparatus called a *zaguri-seishi*, and hurriedly wraps the string over a brass wheel, then feeds it through a stick and finally the reel. Sada starts cranking it. The wooden rod wags like a wind-up metronome as the string clumsily leaves the bobbin. He quickly whips around me, frantic footsteps across the tatami mat, and darts into another closet on the opposite side of the house. Back in a flash, he opens a trash bag with what looks like white and yellow packing peanuts. These are cocoons. Hundreds. Each rattling with a mummified silkworm.

He pulls thin threads from one and then points to the zaguri-seishi, indicating how silk was originally wound. I nod my head. He struggles, trying to explain a factoid, so I bring up my iPhone's translator app. One cocoon string, it appears, can reach the length of "200 mustaches." Later I find it in fact averages 2,000 feet. The mustache conversion table escapes me. He scoops about 40 cocoons into a grocery bag and offers it to me.

"Gift. Gift," he says enthusiastically.

"Dōmo arigatō gozaimashita," I respond, thanking him very much, hopefully not butchering it, and not quite sure how I'll get these potentially disease-carrying pods past TSA security. When I step outside after poking around some, I see Sada has resumed his babysitting. He's smoking a cigarette again, this time visibly satisfied.

Making my way downhill, away from the farmland portion of Gunma Insect World, I see the 3,600-square-foot greenhouse. The giant glass enclosure has an interpretive quarter dome draping over it like the hardened wing of a beetle. Inside is a double deck of semitropical plants from Okinawa. A small bridge passes over the indoor waterfall, and old men with telescopic cameras lean into plants in a full Japanese bow. Going through the sliding door, I arrive at an educational mecca of insect books so high it requires a fireman's ladder. Boys and girls are nudged shoulder to shoulder to gawk and pet furry brown elephant beetles in sawdust terrariums. Giant stag beetle statues cling to the concrete walls. A paper craft table features origami praying mantises made by kids. Educators clasp their gloved hands behind their backs, answering questions. And several boys reach into the beetle boxes and lift the tiny black-shelled tanks to see their intricate undersides.

It's the same exhilaration I see at my next stop: the Jean-Henri Fabre museum. A two-hour train ride from Gunma drops me back off in Tokyo near Nippori's fabric town. Inside the museum, a four-year-old boy rapidly names off beetle species to an entomologist—counting faster than his bungling fingers can keep up with. Aside from showcasing the scientific benefits and anatomy of bugs, the curators have gone to extreme lengths to re-create a room from Fabre's home in the southern part of France—from its aged wooden furniture and door to a window overlooking a hi-res photo of the town.

Outside of this "cabin"—and Japan's animistic reverence of insects—the relationship with bugs takes a drastic turn toward showmanship. The sport of bug fighting* is a long-held Asian tradition. I'm talking about real-life *Mushi-King* beetle-wrestling. One can find multiple venues for summer beetle bouts where elementary school boys pit their pets against each other to play king of the mountain. But China takes the cake for one of the weirdest insect matches.

During the Golden Autumn Cricket Festival in Qibao, just outside of Shanghai, locals and travelers bring crickets in pots and bamboo cages. The festival includes champions, contenders, and underdogs alike, according to *Insectopedia* author Hugh Raffles,

---

*Much like the website JapaneseBugFights.com, which pits various species—like praying mantises and Japanese giant hornets—against each other UFC style.

who gives an incredibly detailed account of the sport. You can find grappling mandibles, quick (six-legged) footwork, and training sessions worthy of a *Rocky* music montage. The chirpy opponents come from select stock, purchased by men with an eye for talent and tenacity. In nearby casinos, pit bosses watch hundreds of men gamble on these warriors as they battle in 60-second rounds. They huddle in sweaty, cigarette-plumed masses around the tiniest arenas with a cacophony of cheers and anguish, placing bets from $1,600 to over $150,000.

Over four weeks, 500,000 gamblers visit China, spending over 300 million yuan, writes Raffles. There are scoreboards, weigh-ins, trainers massaging crickets with grass blades, cricket casino bosses due their percentages, assassination attempts, and even ecstasy "doping." Such matches date back to the seventh century, and the affection for the sport and relationship between man and bug was documented by thirteenth-century chancellor Jia Sidao. Of the five virtues of cricket fighting that Sidao states, my favorite is the third virtue: "Even seriously wounded, [the cricket] will not surrender. This is loyalty."

Collecting crickets and, later, beetles was more of an upper-class hobby, says Eiji Ohya. Eiji, a Japanese entomologist I connected with via listserv Entomo-L's discussion forum, has been kind enough to meet me in Tokyo. More specifically, we're at Nakano Broadway—an epicenter for manga, collectibles, and all things kitsch and anime fandomonium known as *otaku*. We're having lunch together before meeting the manager of Mushi-sha—by far the country's best-known beetle pet store, dealing in both live and dead specimens. I order *unadon*. As I pick away at the fine bones in a grilled eel, Eiji tells me how aristocrats during the Heian period would collect insects for their sound. When Japan was governed by samurai in the seventeenth century, and there was peace, the populace began collecting crickets as well.

Without a doubt, Eiji was a bug boy, a konchu shonen.

"So were 50 percent of the other boys," he tells me. He

reminisces about one summer assignment that required collecting wild insects to display in a shadowbox. (In fact, we later pass by poster boards in front of the OIOI department store building for one such insect scouting project the company organized for children.) Because he was bad at sports, specifically baseball, Eiji would loll about the outskirts of the outfield. "Insects," he says, were his "only friends."

It was during the nineteenth century's Meiji period that shops like Mushi-sha hit the scene. "Mushiya, or insect shops," writes Akito Kawahara, "sold collecting equipment and singing insects in cages." One of the biggest companies, Shiga Kontyu, has been selling supplies (catching nets, rods, carrying cases) to enthusiasts for over 85 years. Mushi-sha began as a mail-order company back in 1971, selling dead bugs through its catalog—one time reaping 1 million yen for a single stag beetle specimen. Starting in 1998, it opened up shop for live beetles. Today it has customers flying in from South Korea, China, and Thailand.

Eiji and I leave the restaurant and walk through the busy streets of Tokyo's Nakano ward, which despite the thousands of people walking around has a quiet to it. And conformity. Even the sidewalks have a dividing line indicating the direction of foot traffic. The nondescript building housing Mushi-sha is near the JR station. The store shares the mildewy hay smell of insectariums. Bags of beetle chow picturing tiny grubs holding forks and knives line the wall near rows and stacks of shiny, barbed beetles sluggishly crawling about their cages. Autographs from local celebrities cover a portion of wall near the ceiling. One is from the pop-rock band Spyair. And waiting in the back for my willing translator Eiji and me is the very reticent store manager Noboyuki Kobayashi.

I ask him how it feels to be operating one of the most successful insect shops in Tokyo. Kobayashi talks for a while as Eiji nods his head and makes subtle murmurs of agreement. (This is actually a form of conversational etiquette known as *aizuchi*.) "For the

Good Times" by Kris Kristofferson plays on their stereo in the background.

"He just likes insects," Eiji laughs. "That's all."

Oy, I think, *this is going to be tough.*

After some cajoling, I learn that they are busiest in July and August, as that's when there are multiple insect festivals and beetle wrestling tournaments throughout Japan. They'll get 300 people coming in a day, lining up outside during their weeklong sales. Kobayashi will work nonstop for 50 days, losing weight as a result. Unsurprisingly, he was also a konchu shonen. As a kid, his favorite beetle was the same one decorating the Tamamushi Shrine.

For your average seven-year-old kid, Kobayashi might recommend an Indonesian stag beetle that goes for 4,320 yen. The Beetle Boom with kids was in 2003 or 2004, he says. Finally my ears perk up when Kobayashi utters the name of the gaming company of my childhood nostalgia, "SEGA." Yes! I know that one!

"Ah, *Mushi-King*!" I interrupt enthusiastically.

The guys listening behind the counter erupt into laughter. I've proved I know more than five Japanese words. Back-pats all around.

"Kids want the real thing," Eiji translates.

One Mushi-sha employee was actually an adviser for the video game. But its popularity is declining, according to Kobayashi. The newer version is intended for the grown-ups who enjoyed the original as a kid (and goofy foreigners traveling halfway across the world). One employee tells me how the enthusiasts are practically entomologists. "They know much more about the insects than *me*," he says via Eiji. "It's a very Japanese phenomenon." Although Mushi-sha's market may wane, given the oscillating trends of pop culture, it's always been more about the message for Kobayashi. I turn to Eiji and ask: What does the manager hope to accomplish?

"By selling insects to kids," Kobayashi says, "they can learn what life is. The importance of life. That they will protect nature and biodiversity . . . Without that," he continues, they wouldn't know "how precious life is." Again, this relates back to animism. "We have many other pets, like cats and dogs, but they are all controlled by the human beings. But insects are more wild and free. Between nature and human being. So, yes," Eiji translates for Kobayashi, "I sell insects. But I want kids to collect bugs by themselves. Go to nature and collect insects and learn. Insects are a sort of mediator between human being and nature."

Today, raising insects, in Japan, has grown in popularity. I think it's for the best. Mushi-sha now works with local farmers. People like Akito Kawahara's friend who sold enough beetles to buy a Ferrari. A scene from *Beetle Queen Conquers Tokyo*, a documentary based on Kawahara's *American Entomologist* article, shows his friend driving around in it.

Due to recent changes in legal policies, Indonesian insect hunters are now paid better wages, hiking up the cost of insect importation. Such cost increase is good news for black market dealers. "Many private insect collectors," Kawahara writes, "do not have official collecting permits for neighboring countries." He goes on to refer to beetles as "black diamonds" sometimes stolen from residents and stores. One such pilfered haul was valued at $67,000. And where conservation laws prohibit overcollecting, a strange variety of thug emerges from the shadows.

In 2008, Virginia resident Wenxiao Jiang was expecting a package; it was never delivered. The problem was it made too much noise. Suspicious postal workers decided to take a quick peek. Inside were 25 beetles from Japan, including a Hercules beetle, which Jiang intended to breed and sell. These beetles could potentially be vectors for diseases or invade crops. The Virginian was swiftly charged by police for not having a permit to own exotic animals.

Ah, the illicit act of smuggling insects into the country. Black market traffickers stick insects in whatever crannies available.

Shoving spiders into your undergarments in hopes of sneaking past customs is ill advised—that's just me—but past endeavors show the benefits outweigh the risks. Take, for example, docile and endangered Mexican redknee tarantulas that can score hundreds of thousands of dollars. Experienced international traffickers make a career dodging USDA fines that reach from $50,000 to $100,000. In 2010, in a government sting called Operation Spider-Man, feds arrested 37-year-old German national Sven Koppler at LAX for smuggling over 300 live Mexican redknees. He had hidden the tarantula spiderlings in drinking straws, possibly making, US agents believe, $300,000 in sales worldwide during his 10-month selling spree.

Koppler is the type of smuggler you might meet at an exotic pet expo. That's why I find myself standing at the bottom of a staircase in an expo line spanning four stories in the Tokyo Metropolitan Trade Center.

For years the exotic pet expo called Blackout was known as the place to buy a large array of insects—like deer-antlered scarabs, Japanese rhinoceros, Hercules beetles—and other creepies. Mysteriously, however, the beetles began to hide from the sales floor. That's not to say these rows of long tables don't offer other exotic pets: sugar gliders, chameleons, owls. Jun Okochi, a Japanese-American gem dealer born in Texas, explains why there's a lack of beetles at Blackout.

It has to do with the man standing on a shoddy stage a couple of inches off the ground under weak fluorescent lights, currently raffling away prizes as sleazy '80s music better heard with vibrant spandex pants plays overhead. His depressed lion's mane hair drapes onto his unbuttoned red silk shirt, which drapes over stylishly frazzled jeans, upon which is a gun holster for his cell phone. This man is Satoru Watanabe—ringmaster of Blackout and formerly Japan's greatest beetle smuggler. He also spent a period of his life as a rock musician, dressing in theatric anime-like band costumes, a genre known as *vijuara kei*. (Please look this up.)

Jun introduces me to Watanabe as he strides through the hall talking to his assistant, but the meeting is a brief handshake and business card exchange. Having not encountered many Japanese people fluent in English, I ask if Jun would like to meet for drinks and tell me more about Watanabe, whom he's known for a couple years. A man he calls "Beetle Master."

Two days before I leave the country, Jun and I commiserate over beers in a Shinjuku bar.

"So," I start, "tell me about Satoru."

"What d'you wanna hear?"

"You told me that he was one of the biggest—"

"Smugglers," he laughs, taking a sip of beer.

Jun explains how the Beetle Master frequented Southeast Asia, paying the locals low wages to find insects for him. After placing his orders, he'd wait for another two to three months back in Japan. "Then he flew back to the country—Indonesia, Philippines, Thailand, wherever—and bought up the insects they caught. And then he smuggled," he says, laughing again.

I tell him about the "Spider-Man" smuggler and the straws he used. "That's how [the Beetle Master] did it until he was caught," says Jun. "But for the beetles it is impossible to put them in straw." Jun's tip for future insect smugglers? Don't put them in your carry-on luggage. "At customs, you say, 'I have nothing to declare,'" he says, his hands innocently raised in the air. "If you're lucky enough, [the TSA] will just let you go through. I hear Satoru was doing it for over eight years. And at last he got caught."

Centipedes, scorpions, hissing cockroaches. From what I could tell, Satoru Watanabe was hauling big time. His operation ran from the mid-2000s to about 2013, but the eventual shakedown happened in an Indonesian airport. Watanabe was smuggling gold beetles. "He was making millions!" Jun continues. Watanabe still keeps mum about how he was busted. Afterward, the beetles at Blackout were slowly phased out.

Jun and I walk through Shinjuku and make our way to Piss Alley—a series of dank, quirky bars and eateries squashed together, each seating no more than 12 people or so. The alleys themselves are narrower than a supermarket aisle. And it's true to its name; after we pass a bar filled with Lucille Ball–inspired crossdressers, we see an old drunk furiously urinating with garden-hose gusto onto the concrete. He stares at us and cackles. So, we cautiously tiptoe around Mr. Peepee and duck into a bar called Albatross, where Victorian chandeliers are a head-bump away. After some plum-infused *umeshu* liqueur we continue down Shinjuku's busy streets. As if I needed reminding we're in Japan, we walk past a life-sized, realistic bust of Godzilla's head over Toho Studios' office (also producers of *Mothra*); the head seems to be peeking over the wall.

This prompts Jun to talk about Japan's eclectic variety of social and cultural differences. A land isolated by water. "I call it '*shima* complex,'" Jun tells me at Zoetrope, a tiny bar playing silent movies and serving the world's best whiskeys. *Shima* means "island." And true to his definition, it's only on this island where such an incredible insect world could develop.

On my flight back home I can't stop thinking about the charming differences of shima complex—bugs aside. The welcoming ear-pop while riding the Shinkansen bullet train. The instant warmth holding a can of Suntory black coffee from a vending machine on a dewy cold morning. The lace-lined taxicab headrests giving the impression the interior was detailed by your grandma. The power-line hum of bees hidden in the artfully sculpted *niwaki* plum blossoms. The subtle jerk of a miniature Honda Kei truck shifting gears through mountain roads, past Akima Shrine. The heightened blood flow plumping your skin from soaking in a tiled *onsen* filled with 111-degree water. The hollow swoosh of the paper *shōji* door. The child grunting "Monkeyyy!" at Jigokudani's snow monkey park (selfie sticks prohibited). The slurpy madness over nasal-dripping steam of ramen bowls, sitting cross-legged at a table with construction workers in blue plainclothes. And the mellow, nighttime pulse of red lights dotted across Tokyo's sky-scrapers—a heart monitor blip until dawn.

❦

At US Customs, I declare my goods: tasty plum sauce and novelty almonds coated in white chocolate to look like cocoons. Never mind the real ones in my checked luggage. But for every couple of smugglers who do exist, there are legit enterprises in this realm of big bug biz: cricket rearing and the legal transport of exotic but-terflies and beetles.

Located in a nondescript building near Denver International Airport is an office filled with tiny white boxes marked with three-letter initials, designated for North American cities. Inside, but-terfly chrysalises hardened like coarse, booger-y spires from across the world are metamorphosing. "Homeland Security hates me," says Richard Cowan, owner of LPS, LLC, and the self-proclaimed "kingpin" of insect trade. These animals would make great bioter-rorism vectors, he tells me, pushing his slippery glasses back above the bridge of his nose. "You can't put a powder in an enve-lope by itself and expect it to work." He's been approached by hob-byists without paperwork selling beetles, but only goes through licensed dealers.

Cowan, who conjures a sturdy Clark Kent in his mid-fifties, has developed a rapport with officials. As a permit-savvy im-porter, he brings in over 1 million special orders a year. Some of those butterflies—for gardens from Indiana's Potawatomi Zoo to the Smithsonian—sell for $3 a head.

LPS has tripled its business in the past five years. His competi-tors are limited, especially now that he's expanded to deliver 200 types of foreign beetles. (One goes for $700.) He even bought a sat-ellite business started by London Pupae Supplies. For the sake of recognition, he kept the company's initials: LPS. Cowan's com-petitors realized the cost-saving benefits of shipping through LPS to get through Customs. Inspection fees after 9/11 went from $50 to $200 a box of insects. "Totaled it's about $500 to get a box . . . so it became, 'Why go overseas for $500 a box when you can go through Rich and pay $50 in postage?'" The process also has to be expedited seeing as the live animals are undergoing metamorphosis. And Cowan has streamlined distribution.

Four hundred thousand butterflies ship through this tiny sealed lab in the middle of strip malls. "Now subtract postage, cost, and everything else, and there's not as much as you'd hope." The other 700,000 insects are in bulk shipping: LPS receives 50 to 80 boxes per week at $50 a box. "Fortunately that profit pays for my butterfly habit," he laughs. Cowan needs it. FedEx delayed a shipment from Malaysia last week, costing him about $10,000 when it got stuck in Fort Worth. "Over half of the box (butterflies, beetles) was dead. [FedEx] didn't compensate me because they don't guarantee the delivery of live animals." That said, be wary around rotten chrysalises. Gases from internal bacteria pressurize the shells. "In this business, we call them 'hand grenades.'" Enjoy washing the stench off when they pop in your hand. It's a problem his two in-office chrysalis sorters experience. When a shipment arrives, they separate the healthy chrysalises from the diseased or poorly emerged ones. When healthy "sleeping bags" break open, their wilted wings take time to dry and flatten. I watch the newly emerged cling to a bookshelf and defecate their last leafy meal, which splats the table like melted Skittles.

The company makes about half a million dollars per year. Yet Cowan takes away an annual salary of maybe $50,000. He is motivated by the challenge of the industry. "When I see a butterfly, I see work." The fruits of that labor are being poured into development of the world's first firefly exhibit—a venture he's undertaken in a secretive lab adjoining LPS. So far it has cost him $150,000. Watching fireflies is an "inexplicable" wonder, he says, that few have been able to witness. The reason exhibits don't exist lies in shipping, and fireflies' 10-day life span. Packaged together, certain

American species of male firefly larvae kill each other like Siamese fighting fish. Cowan's solution was to use a gregarious Taiwanese species of firefly. Eventually his breeding process will enable him to farm 1,000 fireflies weekly.

It may take years before his prototype exhibit is aglow. So far only a few dozen adults have emerged. But nearly every institution he ships butterflies to eagerly awaits his breakthrough and the temporary installment of an exhibit that would be priced at $10,000 a week. The exhibit he hopes to construct will mirror the facade of his fishing cabin in Minnesota.

Even successful insect-rearing operations suffer an occasional crisis. Such was the case in 2010 when a massive outbreak of a virus from Europe killed off hordes of warehouse-raised crickets in North America.

Since the 1950s, Michigan-based wholesaler Top Hat Cricket Farms has mass-produced common house crickets (*Acheta domesticus*) at a rate of 6 million crickets a week for pet stores and lab researchers. You can only imagine their facility's offensively nutty stink. But seven years ago, a Top Hat employee opened one of their incubation bins to a disturbing discovery: the crickets inside had lost their hop. Known as cricket paralysis virus (CPV), the strain spread via the gloved hands of workers, immobilizing their *A. domesticus* breed, turning the sedentary crickets into mush. What's worse was that reports of CPV came from facilities nationwide. The 65-year-old family-run supplier had to trash 30 million crickets, temporarily lay off 30 employees, and opt for a new species, the Jamaican field cricket. But walk into Petco for reptile food today, and you'll be none the wiser.

The outbreak didn't harm the animals ingesting the infected crickets, but CPV managed to infect other cricket farms. Krickets UN Ltd. in Alberta lost 60 million crickets in 10 days, and has closed its doors—as did Lucky Lure Cricket Farm in Florida, a company whose debts reached over $450,000. (Now you can see why the USDA is a hard-ass about insect imports.) The loss put greater demand on Georgia's Armstrong Cricket Farm, one of the largest US facilities, which sells 17 million crickets a week. It remained untouched by CPV.

Large-scale insect rearing holds interesting promises for the future. This last side of insect commerce gives us a glimpse into a new emerging trend—a culinary fascination that begins in cricket farms. There, industrial production methods are inspiring a food source for the twenty-first century onward—a world in which overpopulation will lead to increased food shortages. It's a venture that will include new, young, and innovative millennials. In terms of sustainability, the Old World got it right.

Nine

Dining with Crickets

Drive south on California's I-405, get off at Sherman Way, hang a right, head two blocks into the mini warehouse district, and you'll arrive at Coalo Valley Farms. Unlike your average agricultural landscape, the farm is surrounded by cinder-block walls, garbage bins, shattered beer bottles in the street, and razor-wire fencing with a roll-up dock door. It's a 20-minute walk from Van Nuys Airport—not far from the Laserium where my high school friends and I would smoke weed from cored apples. A group of New Englanders in their twenties, some walking around shirtless, some bearded, tend to Coalo Valley Farms. Their CEO wears a Shih Tzu topknot. But it's in 3,000-square-foot facilities like this that the future food industry is taking shape.

"We're a cricket farm in, uh, the San Fernando Valley," says Peter "Mama Bear" Markoe, the company's chief operating officer. He wipes sweat from beneath a dirty baseball cap as we begin the tour.

Viewed from afar, this setup for rearing crickets—solely for human consumption—is reminiscent of indoor cannabis cultivation. (Even their packaging—small jars and vacuum-sealed foil baggies—reminds me of cannabis products.) Hydroponic Gorilla grow tents, erected as large black cubes, have been "repurposed" into climate-control houses for raising crickets. Water from a koi-fish tank streams into a table containing a shallow rock bed. The flowing water then fertilizes mung beans, their roots sprouting through burlap on top of the rocks. The biomimetic aquaponic system doubles as filtration.

"What we found is it's a great way to recycle water [during California's] drought and organically grow food ourselves," says Markoe. "The fish waste provides all the nutrients needed for the plants to grow in record time."

The fresh spinach, alfalfa, and watercress is fed to about 100,000 crickets, which collectively weigh about 45 pounds alive and 10 pounds once the mature little Jiminys are dried and milled into powder. And then the crickets can be mixed into, say, a protein smoothie—drinks these farmers have prepared for students at the nearby Granda Hills Charter High School, at the Natural History Museum of Los Angeles, and for senior citizens at gardening clubs. "It was cool too," Markoe adds, "because traditionally older folks are not too open to things like that."

Since Coalo's arrival, eating insects—a worldwide practice* known as entomophagy—has made some local impact within the San Fernando Valley.

On the wall hangs a framed 2015 cover image from local magazine *Ventura Blvd*. It features Coalo Valley Farm's tie-dye-shirt-wearing, hiking-sandaled cofounders who crowdfunded the venture via Kickstarter and repurposed the San Fernando Valley location several months earlier. A colorful poster made by UCLA students hangs by some hydroponically grown buckwheat grass

---

* To be more precise, that's 2,086 species of insect eaten by 3,071 different ethnic groups in about 130 different countries.

with painted cartoon bugs and the glittered tenet: "Treat yo self . . . to Insects!" Inspired by the protein-producing operation, the students made a food documentary (and a cappella song) called *Coalo-fornia Dreaming.* These new-age farmers might just convince a small portion of Americans to do what four-fifths of the world already does with nary a wince: eat bugs.

Westerners show a growing interest in entomophagy despite the "yuck" factor. While entomophagy dates back millennia, it wasn't until Vincent M. Holt's 1885, one-shilling manifesto *Why Not Eat Insects?** that someone suggested the modern world cast aside its "deep-rooted public prejudice." Over the next century the occasional publication would extol the merits of the practice, pushing for a more "broad appeal." Examples include Ronald Taylor's 1975 culinary guide *Entertaining with Insects,* Gabriel Martinez's *Cuisine des insectes: À la découverte de l'entomophagie (Insect Cuisine: Discovering Entomophagy),* and Peter Menzel's photo essay *Man Eating Bugs.* But today, some people predict that insects will become engrained in our diet within 10 years.

Since 2009, University of Georgia entomology professor Marianne Shockley has kept a running list of edible insect media mentions. "You went from having a publication once a year to once a month to five times a week," to now sometimes 20 times a day, she tells me over the phone. Why so much interest? As Sonny Ramaswamy, director of the National Institute of Food and Agriculture, once told NPR: "Edible insects are going to be part of the toolkit for us to achieve global food security." In 2010, a student at Princeton began the Environmental Discourses on the Ingestion of Bugs League, aka EDIBL.† Chapters have formed nationwide. Many have joked that crickets are a "gateway bug." "When you ask

---

* The opening dialogue: "Them insecs eats up every blessed green thing that do grow, and us farmers starves." To which there is the reply: "Well, eat *them,* and grow fat!"

† Starting in 1988, a similar group touting entomophagy was begun by University of Wisconsin-Madison professor Gene DeFoliart. Although it was largely for academics, the *Food Insects Newsletter* really began the discussion about insects' nutritional features.

someone, 'Have you heard about eating insects?'" says Shockley, "most likely the answer is going to be *yes*."

The United Nation's Food and Agricultural Organization (FAO) is in large part responsible for entomophagy's growing recognition. In May 2013, the FAO released an extensive report headed by Paul Vantomme entitled "Edible Insects" that promotes reasons for Westerners to embrace this Old World food. One of its authors, Dutch entomologist Arnold van Huis of the Netherlands' Wageningen University, has been a pioneer of eating bugs for the past two decades. It began when Van Huis took a sabbatical in Africa in 1995. There he tried a number of species like termites (which according to one bartender I spoke to taste a lot like strawberry cheesecake). Van Huis then introduced entomophagy to colleague Marcel Dicke and the two began lecturing on the topic.

Their "Edible Insects" paper was downloaded by millions. A year later they published *The Insect Cookbook: Food for a Sustainable Planet*, in which they build on earlier research conducted in their lab, such as their greenhouse gas (GHG) study showing that 18 percent of the world's emissions come from livestock rearing. GHG emissions from cricket and cockroach farms are almost nonexistent. David Gracer, a vocal entomophagy advocate, puts it most eloquently: "Cows and pigs are the SUVs; insects are the bicycles." Bug-sized footprints aside, the FAO report and *The Insect Cookbook* address a larger point: "Eating insects makes us aware of our place in the food chain, of ancient traditions, and blind habits as well as new possibilities."*

In mid-2016, Marianne Shockley and other entomophagists led Eating Insects Detroit. Over 200 people attended the conference

---

*Possibilities Martian colonizers should embrace, according to one 2005 study from Japan. Given the food-regenerating restraints of space habitation, scientists looked to "protein-rich" sources that didn't go *moo*. Out of all the insect cosmonauts—hawkmoths, termites, drugstore beetles—silkworm pupae had the right stuff. Plant pollination, flightlessness, the ability to recycle organic materials, and the essential fatty acids made them prime space food candidates. Also included was a recipe for silkworm cookies.

with the goal of establishing a coalition of like-minded business-people in this uncharted new territory in the food industry. The conference gave birth to the North American Edible Insect Coalition (NAEIC). So far 145 individuals (companies, farmers, curious entrepreneurs) have joined. "It's a bit of a double-edged sword with current [FDA] regulations, which doesn't mention insects," says Shockley. "For some people that makes them a little wary. But it's kind of a saving grace because as long as we're going by all the other protocols, they don't have to mention" that they're processing micro-livestock.*

In a follow-up phone conference, members detailed the bureaucratic steps necessary for the NAEIC and subsequent lobbying that may be needed. Ryan Goldin, cofounder of Entomo Farms, mentioned that "Canada is really stepping up and seeing this as a real industry." A Coalo Valley Farms representative was on the call too.

Food inspectors treat the Coalo warehouse as a general food processing facility, Peter Markoe says, standing in their offices. But Coalo hasn't been producing large enough numbers to merit an FDA approval. "We're exempt in that sense," he says. "It's cool to be on the cutting edge. But it's also scary. There could be this rule that insects *can't* be raised in a warehouse because of a high ceiling or something. Who knows?" Apparently big animal-feed-producing farms (recall chapter 8) are retrofitting their lines for human consumption. Markoe's not too concerned, though, adding "Not too many people want to buy a can of tuna fish from Iams."

To raise their superior, organic, micro-livestock "tuna," the team first tested different cricket species. The *Acheta* family, they learned, was susceptible to cricket paralysis virus. They instead

---

*However, more recently, an FDA spokesperson mentioned that "nutritional companies," a designation that covers many insect foods, don't have to undergo an expensive regulation process that could cost upward of $500,000 and take five years.

opted for the sturdy tropical house cricket. But more experiments lay ahead.

The "baby tent" holds two different sets of humidifiers and space heaters. Unzip the door and the 90 degree ambient temperature and 90 percent humidity radiate on your palm like a warm car engine. The 90-90 formula—discovered in the farm's current v3.5 iteration—gets nymphs through eight molting stages into adulthood. But sometimes the crops don't survive. For instance, a Los Angeles heat wave might dehydrate the crickets to death. Too much water and they drown. "Trial and error," Markoe says, referring to their lost batches. "A lot of trial and error because there's no real book out there—at least for a commercial level."

In another section of the warehouse, the company is experimenting with the mass-rearing of mealworms, similar to what you'd see in a Heinz condiment lab. The room is a Mylar fort with a funhouse mirror ceiling and polyiso insulation boards. A hose for the water-based radiant heating system runs beneath black floorboards, keeping the bugs active. It's another cheap, innovative, green practice to minimize their carbon footprint as much as possible. Sometimes the overhead warehouse lights are turned off due to the excess heat. The Coalo team, all of them friends from Colby College in Maine, miss the cold. But their LA relocation stems from logic. "[Here] you're seeing a cross of Central and South American diets, and also Southeast Asian cultures that [practice entomophagy] as well," says Markoe. "This was a mainland crossroads for that."

He pauses the tour in brief awe of a female cricket, plump and ready for harvest. The adult crickets from one tent will soon be transferred to a clean room where they will be dried and their limbs will be removed in a salad spinner. Same goes for ovipositors. They're the pointy tails on crickets, which, if caught in the throat, could be a bit "rough."

Coalo Valley's largest client right now is energy bar company

Lithic Nutrition. And for Peter Markoe, who followed his co-founding friend Elliot Mermel into this opportunistic sector, eating bugs has been eye-opening. "I had a creepy crawly fear," he admits. "People are afraid of insects. That was me. I've definitely cured myself of that."

☙

A different fear underlies this recent Western push toward entomophagy. Fear in the form of the estimated 9.1 billion mouths to feed by 2050. Food production will have to increase by 70 percent in order to meet the demand of 2050's estimated world population. Added to this are the 100 million children who currently die of malnutrition every year. It's fairly obvious: the need for low-cost, sustainable food sources has become drastically dire.

So why the big bug-eating hoopla? It turns out insects are nutritional nuggets—storehouses of proteins, fibers, and vitamins.

"Developing human brains depend on long-chain essential fatty acids: the omegas-3,-6, and-9," writes Daniella Martin. Her book *Edible*—a culinary travelogue praising bug diets—details an array of economic, anthropological, and delicious reasons why we *should* eat bugs. Martin points out that some researchers theorize termites' long-chain fatty acids helped humans initially evolve. You know those nature shows where apes jam sticks* into logs and pull out termites? That gross practice might be the reason we can do multiplication in our heads.

Crickets are a great source of omega-6 fatty acids. And like mealworms, they are rich in B vitamins. Other bugs have healthy levels of vitamin E and beta-carotene. On a dry weight level, some grasshoppers comprise 60 percent protein, write Marianne Shockley and Aaron Dossey. They also contain 50 more grams of

---

*Chimpanzees will sometimes use their 20-piece "complex toolkit" of sticks etc. to free bugs from trees and crannies like cracking a safe.

protein per kilogram than ground beef. Lucinda Backwell, a pale-ontologist, notes that rump steak provides "322 calories per 100 grams." Termites, however, provide "560 calories per 100 grams." Cold-blooded animals don't need to draw energy from food to keep their body temperature level. So all those extra nutrients are stored away. To find out what that's worth, scientists measure the efficiency of conversion of ingested food. Sheep, with all that cease-less noshing, have a value of 5.3. Caterpillars? You're getting a range of 19 to 31. Another bonus? Diseases like swine flu are only transferred between warm-blooded animals.

Livestock and the production of their feed take up 70 percent of the world's agricultural land. As we saw at Coalo Valley Farms, container bins can store not only Christmas decorations but also large quantities of food. The food conversion ratio (FCR) for beef far outweighs that of insect meat. One pound of beef requires cows to consume 1,000 gallons of water and have two acres to graze, writes Martin. One pound of insects requires a gallon of water and two cubic feet. (The more crowded and orgiastic, the more fruitful.) The FCR for cows—the amount of feed to produce a pound of edible meat—is 10:1. Crickets? 1.5:1. In the end, micro-livestock contain three times as much calcium as beef; and on a dry weight level, mopane caterpillars have 12 times the amount of iron.

There are drawbacks, as with any food. Chitin, the sometimes-waxy exoskeletal coating of all insects, isn't easily digestible and sometimes "limits nutrient absorption." But certain species, like the nutrient-rich black soldier fly, have smaller, more digestible levels of chitin. It's also been "speculated that excessive con-sumption" of exoskeletons, writes food scientist Ruparao Ga-hukar, could lead to kidney stones. But groups like the NAEIC and scientists and economists are not pushing for us to become insectivores. Bugs merely represent a new food ingredient with vast potential. Just skip the all-you-can-eat insect buffet, all right?

The inherent advantages make entomophagy a worthwhile enterprise for food scientists and large industries—industries that could scale up enterprising urban operations like Coalo Valley or Ohio's Big Cricket Farms. But these evangelicals are besmirched by public aversion. Eating insects is often regarded as sci-fi fodder, like entomophagy's negative portrayal in the film *Snowpiercer*. Reintroducing this practice to appetites shaped and informed by culture over millennia remains one of the biggest magic tricks entrepreneurs are trying to figure out.

"The Western abhorrence of eating insects," observed two anthropologists, "is unusual on a global scale." As much as 64 percent of animal protein diets in the Democratic Republic of Congo, for instance, come from insects. Eleven European countries already consume 41 different types of species. For example, in Sardinia, you can munch on the maggots secreting "tears" in *casu marzu* cheese. A common misconception about bug eating is that it takes place only in developing nations. In reality, insect diets are based on what's available in both a geographic and seasonal space. The majority of the West is simply turned off by odd-looking foods.

Neophobia, or the fear of anything new, is the same guiding fear behind a toddler fussily dodging an incoming fork of Brussels sprouts. New foods and their "unknown variables," writes entomophagy expert Julieta Ramos-Elorduy, scare us—and healthier ones almost more so. And it's hard to avoid making contact with the black beady eyes of a cricket before tossing it in your mouth. In terms of looks, though, lobsters are equally repulsive. Crabs look like aliens. And chickens aren't going to win any beauty contests. But dismembering animals behind closed doors and reassembling them anew on a plate makes them, literally, easier to swallow. "The biggest difference," Marianne Shockley says about exotic meals, "is the aesthetics. The average consumer in the US *today* is very disconnected from their food. They do not want to see a whole fish on their plate, they want to see a fish fillet. They don't want to have a cow sitting beside them while they eat a hamburger." The "average consumer," she says, wants "to eat the insect in various products."

Fortunately, the new entomological shakers of the world know this. Just as high-quality, artisanal toast has trended this past decade, so too may entomophagy. San Francisco–based startup Bitty Foods incorporates milled cricket flour into pizza, pancakes, muffins, cardamom cookies, etc., for protein-full, gluten-free demands. Bitty was listed in *Entrepreneur Magazine*'s "100 Brilliant Companies" in 2014.

If cricket protein flour sounds like a futile endeavor, take a look at Chapul. The Utah company, one of the first of its kind on the market, bakes ground crickets into energy bars. (Companies like Poland's Ronzo, which makes cricket protein capsules, are expanding into the body builder world.) Chapul can now be found in stores nationwide. Another cricket powder startup called Chirps converts cricket flour into oven-baked chips in three different flavors: hickory BBQ, aged cheddar, and sea salt. The snack line was crowdfunded by two Harvard graduates around the same time Bitty made its mark. And like the Brooklyn-based Exo, Chirps offers a monthly subscription.

"You can farm all the insects in the world, but if no one's buying them? That's a bigger problem," says Chirps cofounder Laura D'Asaro. "So we decided what we really needed to work on was creating the market . . . We make insects fun and approachable rather than doom or gloom or making it all about environmental statistics."

Laura was able to travel abroad in college while majoring in African studies. While in a Tanzanian street market, she met a woman selling fried caterpillars. She bit into one. The caterpillar tasted like lobster. She loved it. Her question was: "Why are people not eating this?" Her roommate and future business partner Rose

Wang had a similar experience in China. The insects reminded her of fried shrimp. Finally the two of them asked: "How do we get people to *start* eating this?"

Before sourcing insects from Big Cricket Farms, they bought them from Petco. Their fellow students were not enthusiastic about either crickets or mealworms. "Getting people to eat insects is like getting people to eat rocks in that they don't consider insects as a category of food. We see them as pests." Bugs needed a bit of a mind bend. Given America's love for the potato chip, the answer was obvious. They got away from "ooey gooey" and gave insects a "crunch." The mouthfeel aesthetic worked.

"I think about walking into a restaurant and being able to order a chicken burger, a beef burger, or what might be called an 'ento' burger," says Laura. The movement is more about slow integration.

As Dana Goodyear pointed out in her *New Yorker* piece "Grub," this recent craze bears a resemblance to one popular food's history: sushi in America. Convincing our culture to eat raw fish was an almost insurmountable crusade until it was richly priced as a delicacy. Sushi migrated over in the 1960s and gained cultural acceptance only 20 years later. Now it's weird if you haven't tried it.

"Baby steps" is the philosophy on which the founders of London's Ento operate, serving bugs to unsuspecting eaters by disguising them as colorful sushi. Take their concept for Honey Caterpillar Roll, which involves flattening fried waxworms like a thin omelet and wrapping it around radish, cucumber, and carrots. That redirection worked. Run by four people also in their mid-twenties, Ento organized pop-up restaurants that were successful in their own right, selling jars of cricket and caterpillar pâtés—taking something "weird and scary," Fraser said in a *Wired* story I wrote, and "chang[ing] their expectations." New Zealand restaurant Vault 21 does this for locusts by cleverly calling them "sky prawn."

But before we get behind the kitchen doors of top chefs defeating our neophobia with creative dishes, I wanted to educate the diner in me that, like most of you, views insects as a novelty food. Here's the thing: The entomophagists I met up with *enjoy* eating bugs live or dead. Me, not so much. Yeah, I've had my share of stunt dining—mealworm-garnished steak, cricket guacamole hors d'oeuvres, baked scorpion garlic bread—but it was all for bucket list notches and Instagram.

I sought a visceral encounter with what others put on their plate. In its naked form. Largely untampered with. Digested, hopefully, with a cast iron stomach.

Flashback: when I leave Japan, I bring several souvenirs back home in my anatomic carry-on luggage, aka my stomach. This is thanks to a Bug Crawl I partook in with two ladies. My entotarian host is a Tokyo resident who goes by Mushimoiselle Giriko—a combination of *mushi* ("bug") and French formality. At Tokyo's Takadanobaba station, I spot her amid the rushing businesspeople thanks to the black beetle mandibles jutting from her hat. Standing beside Giriko is her friend and fellow entomophagist Eri Sasayama, who'll be acting as our translator today.

How I came to know of Giriko escapes me. But while researching for my trip to Asia, I did read a *Japan Times* news article entitled "Waiter . . . There's a Bug in My Soup." It describes the blossoming Tokyo Bug-Eating Club led by head entomophagist Shoichi Uchiyama. He offers delicacies such as Dubia cockroaches—legs trimmed, abdomen disemboweled of its fecal-engorged intestines—over steamed rice. He's been cooking insects for nearly 20 years and publishing books like *Fun Insect Cooking* and *The Edible Insect Handbook*. Additionally there's the 2015 documentary *Mushikui* in which Shoichi eats a raw Japanese hornet from the hive. Mushimoiselle Giriko, who makes a cameo in the film, seems to have achieved micro-fame in Japan over the past 10 years, writing her own entotarian guide, *Eating Bug Notebook*. She can be seen in the documentary catching wild grasshoppers on a riverbank and then frying them during a picnic, dipping them in tempura sauce.

The three of us sit at a small table in a hidden Burmese restaurant called Nong Inlay, waiting for an ento dish—our first on this Bug Crawl through the eccentric neighborhood of Shinjuku.

"We have a game we play where we eat food that appears in comic books," Giriko tells me. One dish was drawn by famous horror manga artist Kazoo Umezu. "In his comic, one girl eats rice with cockroach," says Giriko via her friend Eri. As far as personal preferences go, Giriko recommends Madagascar hissing roaches. But the recommendation comes with a culinary caveat. For this next part in our conversation, Eri laughs, apologizes, and whips out her pocket translator. "We clean out the internal organs," Giriko says. This is a culinary act compared to de-pooping shrimp because "they smell awful."

Before our waitress appears, I examine a list that Giriko has printed out for me beforehand. On it are 13 different pictures of bugs, each with their Japanese and English names: scorpion, ants, locust, silkworm, scarab and diving beetles, spider, crickets, etc. Of course, it doesn't dawn on me that this is not merely a list, but this evening's menu. A brief description accompanies each. For bee larvae she's written: "In Japan, it's common for us to boil them with sugar. We can buy frozen black wasp pupae in Nagano." The prefecture Giriko speaks of is an inland region with a traditional connection to entomophagy, as sea creatures were harder to obtain there. The rest of Japan thinks eating insects is "weird."

"Our diet is becoming very Americanized," says Giriko.

"It's very clean in Tokyo," Eri chimes in. "Many people are not accustomed to seeing bugs or insects," she continues. "Yes. They hate those animals. But in the rural areas, like the mountainside of Nagano Prefecture, they don't hesitate to eat bugs as food. Because from when they are child, there are so many bugs and insects around them."

Our plates arrive. Appetizer portions of house crickets and bamboo caterpillars stare back at us with their glassy eyes. We first munch away on the crickets. Unlike the ones at Coalo Valley Farms, these retain all their limbs and antennae. No guacamole this time.

"They are like potato," says Giriko. I find the taste more like a nut. It's a welcome familiarity. In the next dish, albino worms cluster in a cute, Micro Machine way. Fried,* the normally squirmy body stiffens into a line with pin-sized legs that could be attached to an electrical circuit board. The caterpillars look relatively harmless enough to be packaged on a convenience store rack: Doritos' Nacho Average Bug. Nuttier Butter. Or, for the British take on crisps called Walkers, Crawlers.

A caterpillar crumbles in my mouth. I cough instantly.

"It's a bit dry," I manage to say, clearing caterpillar dust from my throat. I take several swigs of beer. The caterpillars taste exactly like French fries. The scratchy sensation depends on the freshness of the bugs. Or lack thereof, in this case. Even insects have a shelf life. Since the caterpillars have been dead too long, they make for a dry snack. Entomophagy health fears largely coincide with a standard Western diet. Check the sell-by date. Aside from that, Mushimoiselle Giriko brought up another concern. Would her entotarianism harm her baby while she was pregnant? It was a first for her OB/GYN, who ultimately permitted her to fulfill her entomological cravings. By now, Giriko may have started her toddler on insects.

Our next stop takes us into an alleyway by the JR station. In a

---

* And this is how entomologists suggest you prepare bugs. "Some insects serve as vectors or intermediate hosts for vertebrate pathogens," writes FAO adviser Ruparao Gahukar, "such as bacteria, protozoa, viruses and helminths, thereby increasing the risk of disease transmission to humans." Moral of the tale? Treat insects as you would many other raw produce or meat to avoid ingesting pesticide residue or salmonella. But also choose your species carefully. Gahukar warns budding entotarians of insect misidentification. As one person from Thailand discovered, the blister beetle's canthardin toxin had a deadly aftertaste.

dark corner there are Tibetan prayer flags flying and a Pee-wee Herman–style bicycle parked outside. For a moment, I feel transported back to the exotic tent encampments at the counterculture mecca Burning Man. This is Kometosakasu—a meeting ground for the Tokyo Bug-Eating Club. Daylight from the bedsheet-covered windows lessens as we step inside for our 5:00 p.m. reservation. Large dim lanterns cast meshed shadows along the wall of corrugated roof panels and upcycled wood. We take a seat in the back corner of the house. My eyes slowly adjust as our waitress brings three jugs of "very strong" homemade alcohol, flavored with floating dead animals. Venomous snakes. Dead seahorses. And black ants. It's as though the Evolution Store in New York opened a minibar. I get a shot pour of the black ants and clink glasses with the ladies.

"Kampai!"

Bebop jazz plays overhead. The murky ant-ohol (unforgivable puns permitting) is reportedly good for back pain, which is reminiscent of the medical findings on Hymenoptera I discussed in chapter 7. Speaking of the taxonomy order, our starting bowl of black wasp larvae boiled in soy sauce arrives at the table. Dish No. 1: "It contains a lot of vitamin B," says Giriko, "which is good for beauty." I crack my chopsticks apart and have at it. The wasp larvae, after a couple mouthfuls, have a soft, raisiny texture and are very sweet and a bit creamy, but not in an off-putting way. You could imagine mixing them, as Giriko does, with couscous. In the bowl beside the black wasp are mid-sized locusts also cooked in soy. I'm taken aback entirely. Beyond the sweet, brown glaze there's a fine crunch followed by a light, refreshing herbal taste. My ick factor has gone by the wayside as I hog the bowl. The flavor, I learn, comes from the locusts' diet of rice leaves.

But I'm not sold on the soup.

Dish No. 2: Deflated, shiny brown pods float in the bowl. These are silkworms. "Say hello," Giriko says, laughing as I look into all their shrunken faces. Listening later to my digital recorder, my trepidation is tangible. That dizzying, seasick queasiness grips me.

I take a spoonful and roll it around my mouth. It doesn't help that I can feel the pupae's ridges along my tongue.

"It's like a combination of meat and fish and vegetable," says Eri. "There are some people who don't like to eat silkworm because of its characteristic texture. How do you like it?"

"It tastes the same way a fish store smells," I say, sounding like Ralph Wiggum from *The Simpsons*. Like a koi pond. Or algae. But apparently this soup is better than market-bought cans of silkworms, which are not popular with certain insect connoisseurs due to their lack of freshness.

Dish No. 3 features silkworms, locusts, diving beetle, scarabs, bamboo caterpillars, and a scorpion seemingly posed by Lorenzo Forcella into attack position on a stick with a pineapple cube on the tip.

I spin the kebabed scorpion around to view it from different angles, its armor plates gleaming. One of the more difficult things about eating scorpions is, where do you start? I slowly bite into the crisp pincer. Unlike sea crustaceans, terrestrial invertebrates easily give to our incisors. Scorpions taste kind of like crawfish. The scarab, however, has a disheartening texture that's as brittle and hardened as toffee-coated popcorn and, according to Giriko, tastes like it too. Its body contains elytra, those shell-like wings, which make for a dreadful mouthfeel.

"There's a lot of exoskeleton," I comment while crunching on it. The hard-shelled wings roll in my mouth like press-on nails. "Takes a while to chew."

They bring out No. 4—tamago (egg) sushi with weaver ant pupae, which provide a popping texture and added protein—and No. 5: fried rice sprinkled with ant legs. Viewed aerially, the black legs remind me of my sink after a morning shave. "In this dish," Eri explains, "ants are used as a spice." In China, ants can be used as a salad dressing for a hint of sourness. A few spoonfuls later, my uvula is coated in legs like a prickly pear cacti. My eyes turn red from excessive coughing.

"Do you need some water?" asks Eri.

I give a thumbs-up.

Giriko shouts to the waiter, "*Sumimasen!*"

I make a hocking noise, quickly apologizing.

"Sorry." I blush. "I had to do it." Give me a medal for Best American Ambassador. Returning to the large plate of bugs on Dish No. 3, I study one in particular. This disgusting bug unfortunately reminds me of the Texas body farm.

Eri starts: "This diving bug tastes very"—she searches for the right word—"characteristic. It's an acquired taste and smell." And sight. Its limbs are folded in like landing gear on an airplane. Combined with its smooth top shell, it's more akin to a Vaseline-lubed ovular marble, which is made obvious as I try to pinch it with chopsticks.

"Wow, okay, this one's really slippery," I say as it skates around the plate. "I don't know if my chopstick skills are up to par."

Eri laughs, watching it comically slip into the air. Even for the Japanese, it is difficult, she tells me. Finally I pinch it. My soul sighs. "All right," I coach myself. "Giddyup." My teeth painfully carve through distinct layers, releasing a pimple pop of juice. I know this sounds *very* specific, but it tastes like charred broccolini with parmesan, which in turn is a smell a putrid corpse can also emit. I can see why bullies feed bugs to schoolyard prey. Eri's not too thrilled either as she scrunches her face, adding, "It smells like a drain pipe."

I take a heavy slug of ant-ohol.

Our last destination is a cab drive away through Tokyo's polychromatic metropolis to a windy series of alleyways. The rain-slicked awnings reflect lighted storefront displays printed with Mandarin characters. We stroll through the alleyway labyrinth to Shanghai Xiao-to, a legendary hole in the wall. I've been informed that the Chinese restaurant's off-menu specials include cow penis and dog—actual *bark-bark* dog—which is popular winter fare. "This place is famous among those people who like that," Eri tells me. "And the tarantula is expensive here. It costs 2500 yen." (Roughly $24.)

The owner leads us to the back room. I contort my body through the playhouse-sized restaurant like the Hulk at a tea party, stepping sideways through a doorway, head tilted.

A couple of minutes later our server places two plates on the table. Tarantula and centipede—the type of centipede stunt y tourists dare each other to eat from a snack cart in, say, Beijing's Wangfujing street market. Kebab-ed on a stick, the centipede resembles a series of shields with bent spears jutting from the flanks. The voice of logic clears its ant-leg-coated throat. *You're a total asshole*, it says. I then turn my head to consider the tarantula, charred with bits of unruly hair. Eri Sasayama notices the constipated face I'm making.

"Is that okay?" she asks.

"Sorry," I whimper. "I'm just processing the idea of eating a spider. But I did actually help them reproduce recently," I add. (Recall our dear Claudia from chapter 3.) "So, I don't know."

I opt instead for the other creature, lightly seasoned with paprika. I pick up the kebab. The restaurant chatter comes to a complete halt as all eyes focus on us. Not quite sure it's because of the American on the brink of tears about to eat a centipede or because the chef impaled the centipede with a stick-handle through the mouth, meaning I'm about to chomp down on a calfskin-colored centipede butt. Giriko asks to take photos and starts laughing. Eri's never seen anyone eat this—not even Giriko has tried centipede.

Eri translates for her: "But I think it's a very good experience for you to eat them."

"Anus first?" I ask. "Is that the best way to go?"

I bring the side of its body to my mouth. Its head on my fingertips; its stiff cold legs touching my bottom lip like a drooping bobby pin. My mouth creeks open and its legs prickle my cheek— the discomfort you feel welcoming a shudder-causing stranger into your house. I laugh nervously. My guests look on with cringing smiles. Today's word of the day is *schadenfreude.*

My teeth slowly crunch into the dehydrated thing, my jaws resisting due to the innate voice in my head saying *Just what the fuck are you doing?* Giriko lets out a deeply resonant and astonished "Oh" and then speaks in punctuated English: "Fan-tas-tique."

A squeak emerges from the back of my throat as I rip the tail off. My eyes bulge out of their sockets. It's gray! Gray innards stare back at me within this red hollow reed of skin! Reclaiming my eyeballs from the floor, my jaw does this thing where it moves up and down to turn this gray crusty thing into smaller bits of gray crusty things that are small enough to swallow. Giriko and Eri crack up at the photo they've taken of my face midbite in all its goofball agony.

"Is it good?" inquires Eri.

"I'll say," contemplatively, "not really."

She bravely gives it a taste. "I'm not sure how to express that feeling," she says solemnly. "It tastes very weird."

It is time for Giriko to head home to check on her 10-month-old. Eri and I move on to the tarantula legs. A tad "oily" with a taste of overcharred crab. They were cooked right over a fire, hence the butt goop that has exploded and solidified like cooling lava. "That's tarantula shit, right?" I ask Eri.

"Umm, yes. I think so." We both scowl at it. Neither of us are tempted. And thus concludes the Bug Crawl. But as challenging as the dinner plates were, I come away with a new sense of appreciation for insect foods.

Companies like Chirps, Chapul, or Ento may rely on a marketing strategy that masks the faces of insects to recruit entomophagists. But in the culinary world, I think there's potential in making long-lasting impressions without trickery. "A great dish hits you like a Whip-It," wrote Momofuku chef David Chang. "There's momentary elation, a brief ripple of pure pleasure in the spacetime continuum."

While some insects/ingredients repelled me (at least psychologically), these bugs represent a reinvention of our diet. New culinary discoveries. What's interesting, though, reflecting on my insect dining, is that that "momentary elation" was pretty elusive. Remember the soy-boiled locusts? Awareness of my enjoyment of the dish hit me later as an afterthought. This was exemplified by the few morsels left in the bowl as a social courtesy. It was *that* good. This is perhaps the best solution to attract Westerners to insects: deliciousness.

US archeological digs by David Madsen in 1984 revealed "grasshopper fragments" from 5,000 years ago in caves where Great Basin tribes used them as a food source. The Northern Ute tribe occasionally ground them into a cake traded with settlers for supplies. And it was common for Alaska Athabascan to indulge in the plump maggots of gadflies burrowed in the backs of caribou. They're said to taste "as fine as gooseberries." In Mexico, mealworm dust makes a fantastic tortilla and weaver ant eggs are a "delicacy" called *escamoles*, aka Mexican caviar, writes Dana Goodyear. Today this traditional Aztec dish is in high demand— so much so that multiple smugglers have been caught and jailed time and time again. Same for the Africans smuggling prized mopane caterpillars (zinc, iron, calcium, phosphorus, potassium) into Europe. There, insects are said to be worth their weight in gold. The top-selling nonwood forest product at the Sahakone Dan Xang fresh food market in Laos since 1990 remains insects, especially weaver ant eggs and grasshoppers. Second-most sold are vegetables. The Southeast Asian island of Borneo enjoys rice bugs mashed "with chillies and salt . . . cook[ed] in hollow bamboo stems," writes Ruparao Gahukar. Elsewhere in the world, sickly kids are given "insect biscuits."

So it's weird that this long-standing relationship with insects has been so forgotten by the West. If we are to one day encourage future generations to expand their diets, it'll be the sage-like culinary masters who convince them. Bugs are already popping up in Michelin-rated restaurants.

Crickets are prepared regularly at Noma in Copenhagen.* More impressive, though, is the testing ground Noma's chef René Redzepi started called the Nordic Food Lab.† Inside is a room where foraged ingredients are tested, such as umami-rich garum, fermented with grasshoppers. And Brazil's São Paulo restaurant D.O.M. serves a pineapple dessert topped with a spiky leafcutter ant that has a lemongrass zest. The popular Amazonian ant was a serendipitous discovery made by chef Alex Atala while visiting jungle villages. Manhattan's Black Ant restaurant offers a variety of such seasoned dishes—taking a note from Oaxaca's much-craved *chapulines*—like the grasshopper-crusted shrimp.

Meal preparation can be influential in countries where entomophagy is already engrained in the culture, converting naysayers. When Kenyan economist Monica Ayieko wanted to promote entomophagy in Lake Victoria, where tropical heat causes food shortages, she did so with an insect smorgasbord: mayfly crackers and termite meatloaf to name some. Ultimately the termite sausages won with 65 percent of the subjects favoring new culinary takes as opposed to traditional preparation. "Two participants commented," she writes, that "'insects will always be insects regardless of how they are prepared or served.'"

However, this isn't how our brain is hardwired. We actually eat

---

*Children at private Danish daycare Valby Lillefri were asked how they felt about eating foods that may or may not contain insects. Having chosen the children for "their lack of established experience" with food, the researchers presented five snacks: an apple, baguette, pasta with cricket-infused tomato sauce, salad with apparent whole crickets, and a snail. Based on their repulsed facial expressions, it didn't go well. When asked if they'd consider entomophagy, one kindergartener left the discouraging note that the bugs were "disgusting . . . because they are slimy."

†The team of which ran amok exploring insect tastes in the 2016 documentary *BUGS: A Gastronomic Adventure with Nordic Food Lab*. I strongly advise visiting their BugsFeed .com blog that aggregates recipes, including: silkworm spaghetti and bruschetta, wild herbs with honey and ants, waxworm paella, and a Bee-LT sandwich.

with our eyes. And as famous Seattle bug chef David George Gordon told one reporter: "The stomach always votes last."

Le Cordon Bleu may not offer entomophagy courses just yet. But if it did, Gordon would proctor. His accolades include winning a couple of insect cook-off challenges and experimentation like his Three Bee Salad. It calls for bee adults, pupae, and larvae. His signature dish is the Orthopteran Orzo, comprised of wingless and ovipositor-less cricket nymphs that are "more tender than the grown-ups," and parsley to stave off "cricket-breath." He recommends that the crickets fast for a day so they're fecal-free. Next, separate the necrotic bugs from the group, freeze the lot, rinse them, and heat up the skillet.

Gordon favors a visible approach to serving bugs because it ethically does more justice to the bug. Taste triumphs over all. "If it's saving the planet but tastes like cardboard, no one's going to care," he says.

His gastronomic scruples partly inspired his critically acclaimed trailblazer *The Eat-a-Bug Cookbook* in 1998. It landed him dozens of TV appearances, including a spot cooking alongside Conan O'Brien, and lectures across 32 states and overseas. David George Gordon, however, is not an entomologist. And besides taking several cooking classes, he's no certified master chef. His anthropological interest in entomophagy began in 1996. For a year and a half he learned the nuanced flavors of ento ingredients, spending, for example, two months nailing down variances in one bug. Is there a difference in taste from domestically or wild-reared crickets? If they're frozen and defrosted, does that affect the flavor?

"Mealworms, for instance, have a mushroomy taste," he informs me, "so they're great mixed in with sauces." Not so much on ice cream, as others have tried—"that's not a savvy pairing."

A career high point came in March 2015 when he was asked to cater the annual Explorer's Club dinner at the American Museum of Natural History. The black-tie crowd included Lou Sorkin, the

"Bedbug King of New York" (chapter 5), as well as celebrity scientists like Neil deGrasse Tyson.

"I'll stick to this claim that it was the largest bug[-eating] event in recorded history," Gordon says. A range of delicacies were served buffet style, including deep-fried tarantula, bacon-wrapped Cambodian mole crickets, American cockroach canapés, black and red ants on a log, and his signature, Orthopteran Orzo.

The dishes were well received. "My theory now is to do catered events for Hollywood types or the tech industry," he says. Perhaps then it will trickle down to the point where we can purchase prepackaged lunches at a Safeway or Trader Joe's.

It makes you wonder . . . Will edible insects become luxe, effete plates served in an upper-class din, or a mass-produced food with hormone-injected crickets the size of plump hogs (the lifeblood of radioactive monster movies)? Really, it's hard to say at this point. Gordon divulges the extravagant cost to cater the Explorer's Club dinner: $15,000. That was for the bugs alone. While online businesses have made it easier for entotarians, the high insect prices have fluctuated little. The industry is still too young. Also, importers like Thailand Unique, which sells a variety of exotic absurdities like Bugapoop Tea Bags, often ship ingredients that are "desiccated, crunchy and terrible," says Gordon. Ento connoisseurs are just too few and far between to make a stink about it.

Wild insect hunters* have yet to make a mark in the United States. But something may happen if we eat processed or culinary-grade insects on a regular basis over several years. We may actually

---

* The harvesting of insects concerns conservationists. Whereas some maize fields in Thailand are grown solely for locusts to eat, other regions overharvest insects to near extinction. Over 10 species are listed as threatened in Mexico thanks to restaurants. The Aboriginal preference for witjuti and bardi grubs is endangered by "increased exploitation" by ecotourism and restaurants as well. Proof that every farming practice requires sustainable protocols. And cultural awareness: In Mali, children regularly snack on grasshoppers that in turn snack on cotton fields. However, when Western advisers promoted the use of pesticides, the hoppers diminished, as did the children's protein source, leading to malnutrition in 23 percent of the kids.

be inclined to *raise* insects as some hobby farmers do with chickens. It's not much of a stretch. Again, young entrepreneurs are seizing the opportunity.

Jakub Dzamba, a 30-something Canadian inventor, knows this. His Cricket Reactors—vertical Plexiglas mazes—"maintain superior hygiene levels" for human-consumption-grade crickets and can be kept on your kitchen countertop. The prototype modules, which look like two stacked microwaves, can "grow" a half pound in a month. As a long-term goal for the project, Dzamba wants others to mass-produce crickets as either a hobby or an industrial farm. With 100 square meters of Cricket Reactors, you can yield one ton of meat per year. Currently he's working on scaling reactors up to retrofitted cargo containers ready to produce upon delivery. These 200-square-foot Chirpboxes are capable of yielding 2.7 million crickets on a monthly basis, which translates to about 1,200 pounds in food.

Katharina Unger and Julia Kaisinger are also in-home incubator entrepreneurs. The Austrian duo recently brought the LIVIN Farms Hive to market—a climate-controlled desktop kit that looks like a chic metropolitan apartment complex. The multilayered farm, governed by a number of sensors, separates mealworms by their growth stage. Adult meal beetles lay eggs in the penthouse. As the eggs mature and evolve, they procedurally filter into the trays below, composting kitchen waste in the process, with fans reducing that buggy aroma. A past incubator raised black soldier flies—a choice option given their less chitin-y bodies and knack for food by-product conversion. But mealworms had a broader appeal. (I can attest how well they substitute for peanuts with beer.) Maintained properly, the unit churns out up to 500 grams of mealworms per week. That comes at a $700 price tag. But for an endless generator of food? Not too shabby. It also serves as a microscale example of what industrialists can achieve in food security.

By and large, entomophagy provides one answer to world hunger.

"Instead of trying to get all of America to eat bugs," Gordon says, "focus on the people that are already there." He's had success convincing people, although a mild or "not bad" response won't recruit return customers anytime soon. "What's funny is kids are much more adventurous than their parents. They say, 'Gee, I couldn't get him to eat anything at home, and here he is volunteering to eat a scorpion claw' . . . That's where I think the real change is coming."

University of Georgia professor Marianne Shockley sees this every year toward the conclusion of the Bug Camp she hosts at the university. On the last day they'll have a "bug party," and students will cook mealworm cookies and cricket pizza.

I was hell-bent to put such adolescent curiosity to the test in my own group of friends. Would it be possible to not only pique their interests, but potentially convert them? And how much smaller would my small group of friends become afterward? To find out, I created a Facebook invite to a most unnatural dinner for a bunch of folks I've known since high school.

I'm staying, while I visit, in my old room in Southern California's Granada Hills. I'm writing in the now-converted office and miss the doorbell when the delivery man comes. My mom is less than ecstatic when a University of Georgia package arrives on her porch from Marianne Shockley. Among other insect products, she's sent me about a pound of dry-roasted crickets. My mom can barely look at me as I bring it in and empty the contents into the freezer. I can only imagine her face when the live waxworms arrive.

This bug soiree came about with the help of Marianne and David George Gordon.

```
Me: Hi Marianne . . . Hope you'll remember
    me from the ESA conference. I plan on
    cooking up a bug feast for my friends in
```

L.A. I've just narrowed down the dates.
Just wanted to confirm you'd still be able
to send along some ingredients.

Marianne: Sounds great. This is very exciting news!

Me: Here's what we have on the menu (taken
from *Edible* and the *Eat-a-Bug Cookbook*)
as well as the insect quantity for
each . . .

Wax Moth Tacos—1 cup of waxworms
Curried Termite Stew—20 winged productive
   termites
Cockroach Samosas—24 American cockroaches
Three Bee Salad—40 frozen adult bees, 60
   frozen bee pupae, 60 frozen bee larvae,
   1 oz. bee pollen granules
Superworm Tempura w/ Plum Dipping Sauce—24
   frozen superworms
Sweet 'n' Spicy Summer June Bugs—???
Pizza with Cicadas—8 subadult periodical
   cicadas
Chocolate Cricket Torte—1 cup of crickets

Marianne: I'm reaching out to my dear friend
and colleague David George Gordon (DGG) to
help me find a line of these items. Hon-
estly, this is a very small order, and we
have to get their minimum on some of these
items, but I'll certainly let you
know . . . DGG, any thoughts on the termites, roaches, bee larvae, June bugs and
cicadas?

DGG: Thanks for getting in touch. Nice to
see that you'll be using a bunch of my

> recipes. I think the real go-to guy for
> edible insect supplies is Dave Gracer in
> Providence, Rhode Island. If you haven't
> done so, I encourage you to get in touch
> with him ASAP.
>
> DGG (cont'd): Regarding June bugs, I'd get
> in touch with Paul Landkamer of Missouri
> Entomophagy . . . In the past, I've pur-
> chased American cockroaches from Carolina
> Biological Supply. They're kinda pricey,
> but at least you know they were raised in
> a sanitary environment.

Given time constraints (and perhaps it is too much to ask my friends to eat cockroach samosas), I end up serving Wax Moth Tacos and cricket cake. As recommended by *Edible* author Daniella Martin, I go through San Diego Wax Worm and get them shipped overnight. With both crickets and euthanized waxworms in my parents' garage freezer, bug night is under way.

Around 5:00 p.m., I start by chopping hardened crickets. The hollowed, toasty bodies crumble into halves and quarters. Traditionally, in the past, good ol' mama might be sitting where I am prepping meals with our family's throwback recipes from Egypt—grape leaves, hummus, bamya matboukha, molokhiya. But she won't come out of her bedroom until the insects are out of the house.

Score one for the neophobes.

My younger sister Kristen agrees to help me bake the chocolate torte. She shares relatively the same viewpoint as our mom so I am surprised by her cool when I point out the bowl of chopped and roasted crickets I'm about to mix into the cake batter. In this processed form, you can't tell what it is.

Jim Grude and his wife, Tara, are the first to arrive. Two more couples are followed by Matt Nyby and his brother Josh, and later

Nick Gutierrez, my entomologist friend who ignited this recent insect fascination of mine. He's brought guests of his own in a cookie container—a discoid and Madagascar hissing cockroach. To prevent a likely "hissy fit" on the part of my mother—still locked safely in her bedroom—he returns the roaches to the emptied plastic box. Aside from Nick, nobody has ever eaten insects, and I've double-checked to see if my guests have shellfish allergies.*

Drinks in hand, the crowd hovers around the snack table, which contains two bowls of salsa, a vegetable tray, and two large bowls of chips (sea salt and BBQ flavored). Kristen pops the springform pan into the oven to bake for 40 minutes at 350 degrees Fahrenheit. Time to prep the waxworms for the skillet. I take the San Diego delivery out of the freezer. In the blue pots the worms appear as bits of lifeless flesh shaved from a cheese grater. The problem is the worms are mixed in with sawdust. Jim helps me sift a thousand larvae from the four cups of sawdust, picking out the black and graying ones as we go along. Kristen and the others continue eating chips with nary a wince.

"What do they feel like?" Kristen asks. I offer the bowl of larvae and she touches one for a hundredth of a second. "Augh!" She cringes.

"When's grub?" asks the puntastic Matt Nyby. It's nearly 7:00 p.m. now. My guests are starting to get antsy (hey-ooo!) and the larvae are no longer cold. I get the onions golden in the pan and scoop in the mound of worms. They sizzle. I keep stirring them around on the stovetop over medium-high heat. The worms straighten and glisten. Their outlines turn transparent. I slide them into a serving bowl and place it alongside cilantro, tomatoes, tortillas, and cheese.

---

*Insect allergen sensitivity is a rarity for those who haven't been exposed to bugs on the reg. But some have asthmatic responses to *Orthoptera*. Others react poorly to waterbugs. One study found those overexposed to dust mites grew "sensitive to seafood tropomyosins." And by overexposure, I mean people who regularly breathe in cockroach-crap-infested dust particles.

But the novelty aspect of the tacos is not the real test tonight. After arranging the taco line, I turn to Kristen. "So, you ready to eat some bugs?"

"I might have the cake," she says with a suspicious glance.

"But," I start, "at least try the tacos, you've already eaten crickets." She stops in her tracks. "Those chips," I say, her face reddening, eyes wide and mouth pursed, "are made from cricket powder."

Kristen is dumbfounded. "What?! No!" But once the initial shock subsides, the curiosity sets in. A couple of friends thought it was a "healthy" chip—the ones you'd find at higher-end markets, made with flaxseed or wheat. People continue eating them, and Tara asks where I got them, taking a picture of the Chirps bag I bring out. I ask everyone later how it makes them feel. Overall, there's interest. It is healthy. And maybe, were it widely available, they might pick up insect-derived processed foods. Nick is "elated" to find out they were made of crickets. But my sister remains opposed for now.

I'm happy to say the tacos and cricket torte were a hit (and that Kristen bears no grudge for the dupery). Jim Grude proceeded to eat three tacos. And the overall mood of the crowd was one of titillation. It reminds me of something Marianne Shockley said during our phone conversation. She teaches about 250 college students every year. "Very rarely does a student finish my class without having tasted an insect. Are they going to go out and buy something the next day? Maybe. Maybe not. But do they at least overcome that initial barrier? Yes." Over 15 years of teaching, she's converted thousands, possibly tens of thousands, into entomophagists. "My mission . . . is really to let people know how diverse, how beautiful [insects are], and how they, literally, *are* all around us. All the time."

Humans and insects. I'm not certain where our bug affairs will lead over the next couple centuries. But the optimistic cynic in me likes to believe that it might reflect one of the longest marriages we've had with nature. One that requires going back to what is arguably the most imperative species on Earth, depicted on cave walls in Valencia, Spain, 15,000 years ago: honeybees.

Ten

Tracing the Collapse

The glass door on Logan Street reflects the empty silhouette of a bald man with two large silver earrings. I buzz Lawrence Forcella, aka Lorenzo, aka the bug-pinning artisan from chapter 1, into my building.

Lorenzo's taking a break from his eight-hour drive from Wyoming after scavenging for dead animal bones and picking up an antelope head to sell to the Evolution Store—a collection of oddities that has relocated from New York's SoHo neighborhood to NoHo, closing down the Entomology Room I visited in the process. It's hard for him to lose the department he started. Now he pins bugs for his own God of Insects enterprise with the occasional Evolution consultation. When I heard Lorenzo was visiting family in Colorado, I asked if he'd make a pit stop in Denver and sample this local mead I'd just bought.

Back in my apartment, I pour two glasses of black raspberry mead. We toast and fill each other in on the happenings since I last saw him. I introduce him to Bill "Fucking" Murray sitting on my kitchen countertop, my discoid cockroach/botched attempt at cyborg science and current house pet. "You know what the selling point on the *Blaberus discoidalis* is?" he asks me. I wait for the punchline. "They like to disco." The two nerds in the room start laughing. He gives me tips on how to pin Archy and Bill "Fucking" Murray when they reach the pearly gates, like creating a Jacuzzi for them to soak overnight, making their limbs flexible.

I tell him about Brazil and my visit to a couple of the University of São Paulo's bee colonies. Unlike at your standard apiary, we weren't required to wear protective garb. The variety of bee genus there is known as *Melipona*, and they are stingless. A student assistant lifted the lid off one hive no larger than a jewelry box, and I dipped a taster spoon into the colony's honey bulbs, which looked like a string of quail eggs. The by-product was runny pancake syrup. The taste was invigorating. Wide commercialization of this honey,* though, is not possible due to stingless bees' slow production and restricted importation.

The conversation drifts to honey production in the United States and the dark effects of big agriculture and pervasive toxic chemicals.

"It's a little like saying that any one particular thing is responsible for colony collapse disorder [CCD]," says Lorenzo in that emphatic Brooklyn accent. CCD caused a great stir starting in 2006. You know it as the phenomenon in which 30 percent of honeybee colonies were abandoned, leaving only some workers and queen bees, rendering hives, as beekeeping businessman Dave Hackenberg says, "ghost towns." Beekeepers across 36 states lost as many as 60 percent of their hives in the 2000s. Cases of the mysterious disease have dropped in the past six years. The causality, still very much debated, has been pinned on a number of pests, fungicides, pathogens, and poor nutrition.

---

*Honey known as *tiúba* and *uruçú* in Portuguese, which fetches ten times more than your average jar of the sweet stuff.

I ask Lorenzo, "Haven't massive bee die-offs been an ongoing phenomena through centuries with different names?"

"It's possible, but I actually don't know much about honeybees. No one really asked me about bees until people came up to me asking if I've heard of CCD. 'Well, how the *hell* have you not heard about it?'" he says, feigning aggressiveness. "People have gifted me beekeeping guides that are almost 100 years old. And it has a *long* list of *stuff* that can kill your bees! It's like, oh my God, these things are prone to so many different diseases. And weird disorders that have strange names that obviously no one knew what it was."

"Israeli acute paralysis virus. American foulbrood," I say. (Both diseases are explained in the pages to come.)

"Back then no one studied this stuff scientifically. It was just knowledge passed on from beekeeper to beekeeper."

"So what solution is there for the honeybee?"

It's the pervading question thousands of bee experts, aka melittologists, have sought to answer about the European honeybee (*Apis mellifera*). Lorenzo mentions how there's a "mad scramble" by melittologists to discover native bumble bees before they go extinct. Bees appear on every continent but Antarctica, and have an estimated worldwide value of $153 billion a year. Entomologist Marlene Zuk points out how "218,000 of the world's 250,000 flowering plants, including 80 percent of the world's species of food plants, rely on pollinators, mainly insects, for reproduction." Managed bees pollinate 100 crops in the United States, from watermelon fields in Florida to California's 1.1 million acres of almonds. So it's vital to find out the *why*. Why this decline? And what can we do? Bugs are nearly as necessary to humans as breathing.

I think back on my earlier question at the start of this book: what is a bug? Coming away from this journey, I'd say they are saviors and survivors, shaped by the world to, in turn, shape it.

Our closest tie to bees, and therefore nature, predates the

invention of alcohol—a honey-fermented drink akin to the mead Lorenzo and I drank the night of our reunion. These omnipresent "flying dust mops," bee researcher Jerry Bromenshenk once called them, pockmark all of history. Bees have inspired jubilation, romantic verse, cultivation, medical advancement, industries, and more.

The first solid evidence of the world's greatest pollinator dates to 65 to 70 million years ago, and was found in modern-day New Jersey. Suspended in amber from the Late Cretaceous period is a eusocial bee, part of the classification family known as Apinae, which includes what are called corbiculate bees. Corbiculates include honeybees, stingless bees from Brazil, and solitary bees that live alone in holes, mostly in the ground or trees. But what makes them so important is that their legs are pollen baskets— saddles they use to carry a flower's sperm, picked up with their front legs. They then travel at the hyper pace of 15 miles per hour. While evidence shows that pollinating flowers surfaced 100 million years ago, during the Cretaceous period, they only truly proliferate when bees are present. By calibrating fossil evidence to reconstruct the history of corbiculate bees, Cornell researchers recently hypothesized that bees' eusociality (the ability to nest in colonies, store honey, etc.) may go as far back as 87 million years. Other US scientists have found evidence to place bees even earlier in the Jurassic period, or 200 million years ago.

This creature paves the way for early man. Honey was the original sweetener. Our first interactions with bees are depicted in Spain's Cave of the Spider. The 15,000-year-old drawing shows an androgynous figure hanging by a vine and possibly holding a smoking bush to calm bees. You see, early man was a thief. Bee colonies hid within rock crevices. Dangled off trees. Before managing them with wicker baskets coated with mud, honey hunters

risked their lives (and countless stings) probing around and chopping at branches or cutting portions of single-comb nests* the size
of Stegosaurus plates.

Later on, pottery vessels, present since 5000 BCE of the Neo-
lithic period, were used to store colonies—and enable less harmful
means of consumption. Honey cylinders could be found in southern Mesopotamia as early as 2450 BCE. Here, one preventive measure to deter moths (which produce the comb-eating larvae) from
hives was to "sprinkle fresh milk and the urine of children" over
them, writes *The Sacred Bee* author Hilda Ransome.

The first evidence of beekeeping is found in ancient Egypt. Bas-
relief and hieroglyphs in the Sun Temple of Niuserre from 2400
BCE spell out how humans would be found "blowing or smoking,
filling, pressing" the hives and "sealing honey," Ransome explains.
Eighteenth-century foreign travelers noted the migratory patterns
of Egyptian beekeepers, traversing the Nile in rafts with pyramids
of such potted colonies, gathering pollen and nectar from across
the nation. Such migratory hives are also reported in Rome on the
river Po by Pliny the Elder, with similar accounts from Spain.

Honey soon became an integral part of our culture.

Canopic jars that stored the intestines of deceased Egyptians
had bees carved on them, symbolic of Horus's sons. Egyptian mat-
rimony promised "twelve jars of honey" to brides. Croatian wed-
ding processions required honey smeared on doorposts as brides
were fed a spoonful of honey to "live peaceably" with their grooms.
That's *three* spoonfuls if you live in Serbia. In India, traditional
Hindi beekeepers carry holy basil while tending to the hives—a
practice associated with honey ("food of the gods," as Ransome
points out) and Krishna's maiden. Kama—the Hindu version of

---

*This death-defying stunt is still practiced in Nepal. Gurung tribal members use tradi-
tional tools—bamboo pole rafters—and rope ladders in the biannual tradition of collecting honey from *A. dorsata* bees on steep cliff faces, putting the best acrobatic
performances to shame.

Cupid—is followed by a kite's tail of bees. Passages from the Sata-patha Brahmana place a heavy distinction on honey, calling it the "life sap." Hindi newborns receive a drop of honey.

Soon honey laws became a necessity. Hittite laws circa the thir-teenth century BCE deterred would-be thieves: "If anyone steals two or three beehouses, he formerly had to have his own hives destroyed, now he needs to pay 6 shekels of silver." This is the ear-liest evidence of such legal action. "Bee-Judgements" in Ireland, starting in the eighth century CE, instructed honey payments to tribal chiefs from a swarming hive. A hundred years later, King Alfred demanded bells be rung at the sight of a swarm so it could be hived.

There is plenty of bee superstition. Counting hives in France was bad luck, and stings were thought to be the result of swearing, which really would only beget a cycle of curses and stings in equal measure. In Rome, where slaves and nobles partook in beekeep-ing, the bee goddess Mellona is portrayed with a beehive-topped* staff. Dionysus, a mythological being of pleasure, is rumored to originally have been the god of mead rather than wine. At the time, "bees flying through the cracks of rocks," writes *Sweetness & Light* author Hattie Ellis, "were thought to be souls emerging from the underworld." In Iran, fifth-century[†] BCE historian Herodotus notes, dead kings were enveloped in wax.

Apiculture, or the maintenance of honeybees, was derailed for some time in Europe because of ravaging Barbarian invaders who

---

*Which spurred my curiosity for the beehive-topped head. The 1960s hairdo much be-loved by film director John Waters known as the beehive was invented by Chicago beautician Margaret Heldt. This fez-inspired design, comb-twirled and solidified with aerosol spray, received its name from a reporter observing the bee-shaped pin acces-sorizing the coiffure.

[†] Around this time in 401 BCE, historian and soldier Xenophon and his army retreated from Turkey. The culprit? Debilitation by the country's infamous "mad honey" foraged from *Rhododendron ponticum* flowers. The nectar contains grayanotoxin, which led to a number of symptoms. "[They] went off their heads," Xenophon observed of his soldiers, and suffered from vomiting and diarrhea . . . those who had eaten a great deal seemed like crazy . . . dying men."

destroyed the land. But beekeeping picked up once again as wax candles were sought out by Catholics.

Similar to earlier sanctioned bee laws, "England's Charter of the Forests in 1225," writes Hattie Ellis, "established that taking someone else's honey and beeswax was an act of poaching." (Speaking of minding one's own beeswax, the childish taunt stems from olden days when women smeared wax on their face. If standing too close to a fire, UC Davis entomologist Elina Niño told me, they were quickly warned to "mind" it.) Seventeenth-century reverend and bee fanatic Charles Butler praised beeswax candles, saying that they "maketh the most excellent light, fit for the eyes of the most excellent; for cleernesse, sweetnesse, neatnesse, to be preferred before all other." (Apparently he's never been to Yankee Candle.)

Even with all this bee hubbub, it wasn't until the 1500s that we sought ways to harvest honey without killing the bees.

Before the nineteenth century, beekeepers had a more rough-hewn mentality. The standard enclosure used by Chinese bee-keepers was a mud-coated basket. Hives in Europe might consist of oak cork or wicker baskets or logs originally strung up in trees and then later placed on the ground. German bee masters, aka *bienenvater*, refer to these hives as *klotzbeute*. At times, they were topped by gables, giving the impression that Keebler elves lived inside.* Superstition in western Germany prevents traveling bee-keepers from strapping their hives onto themselves like backpacks facing behind them, lest their bees flee. A colony may also depart

---

* An ancient prototype compared to today's modern alternative hives. For instance, Flow Hive—crowdfunded by a father-son team in Australia in 2015—fits plastic honeycombs on plates that split vertically when cranked like a music box. Influenced by beer taps, the honey flows directly from the hive. BeePak, on the other hand, offers a compartmentalized Samsonite hive weighing less than half of the traditional ones and is built to withstand charging elephants. And for the health nuts out there, mushroom expert Paul Stamets has prototyped a hive composed of compressed sawdust and mycelium. The reason being that *Metarhizium anisopliae* fungi helps eliminate bee-killing varroa mites.

when a bienenvater dies. Similar to traditions in other regions of the world, when a beekeeper passes away, relatives or friends must then tell the hives their master is gone or else "the bees will follow [them] and die." Such animistic lamenting is captured beautifully in this two-stanza excerpt from nineteenth-century poet John Greenleaf Whittier's "Telling the Bees":

> *Before them, under the garden wall,*
> *Forward and back,*
> *Went drearily singing the chore-girl small,*
> *Draping each hive with a shred of black.*
>
> *Trembling, I listened: the summer sun*
> *Had the chill of snow;*
> *For I knew she was telling the bees of one*
> *Gone on the journey we all must go!*

Honeybees arrived in Virginia around 1622. As Thomas Jefferson noted, "The bees have generally extended themselves into the country a little in advance of the settlers." He also mentions that Native Americans referred to them as the "white man's fly," although Central and South Americans began beekeeping before the arrival of Spanish conquistadores. For their trip across the Atlantic, bees were transferred to inverted straw baskets called skeps. Barrels of hives over ice would keep bees calm while traversing eighteenth-century seas. Sometimes anxious, frenzied sailors heaved hives over the side, blaming poor weather on the bees.

In 1657, English cleric Samuel Purchas wrote what might be considered the first beekeeping handbook, *A Theatre of Politicall Flying Insects: Wherein Especially the Nature, the Worth, the Work, the Wonder, and the Manner of Right-Ordering of the Bee, Is Discovered and Described.** At that time, observation hives, with lone-standing

---

* No one ever accused Purchas of being succinct.

window frames, made it possible to see how honeycombs were constructed. By the eighteenth century,* blind scientist François Huber published *New Observations on the Bee*, which discussed queen bee mating and birth, temperature control, and oviposition. He described royal jelly, the foodstuff of queen larvae. But the design of the beehive frames prevented critical research on the inner workings as combs attached to the walls. Destroying them was regular practice up until about 1851, courtesy of a three-eighths-inch revolution.

After observing a friend's glass globe beehive in the 1840s, Massachusetts pastor Lorenzo Langstroth became intrigued by bees. Combs had long been cut into pieces. However, Lorenzo knew of the Greeks' movable combs and how some European hives had wood frames. Combining these ideas, he made slidable frames in a wooden box, removable like circuit boards in a mainframe, spaced three-eighths of an inch—just enough for bees to move through but prevent honeycomb buildup. This "bee space" would change apiculture. Some people must have been apprehensive at first. To them, he wrote, "Those who object to this as interfering with nature, should remember that the bee is not in a state of nature." While his patent for it was accepted, other companies and apiarists were quick to make slight adjustments to their own inventions, echoing Langstroth's bee space. But in the midst of feeling ill and combating depression, Langstroth didn't pursue the lawsuit he'd filed. He died in 1895.

Apiculture flourished. Newspapers reported on the potential wealth to be made, sending flocks of people to California with "bee fever." Those West Coast apiarists would very soon produce 2 million pounds of honey annually, writes Hattie Ellis.

What followed was a result of the bee craze. *American Bee*

---

* At this time in England, Thomas Wildman, aka the "Barnum of Beekeepers," had been demonstrating the harmlessness of bees in public shows by covering his body with them. His most famous trick, replicated by others, started by taking the queen and positioning her on his chin, forcing the colony to form a buzzy ZZ Top beard.

Journal was established in 1861. Honey extraction improved with the use of centrifugal force to spin the sweetener from comb cells. Five years later, in 1870, a New Yorker invented the modern bee smoker. The first train car full of honeycombs was shipped to Chicago in 1873. In 1901, author Maurice Maeterlinck's book *The Life of the Bee* sparked beekeeping fever across the world. Soon after, our nation's first apiculture expert was appointed by the Bureau of Entomology. And by 1909, the United States had its first commercial bee rental.

From the Gospel of Maurice (Maeterlinck, that is): "They are the soul of the summer, the clock whose dial records the moments of plenty . . . the song of the slumberous, languid air; and their flight is the token, the sure and melodious note, of all the myriad fragile joys that are born in the heat and dwell in the sunshine."

Due to a market surplus of honey, we moved on to different extracts. Starting in 1930, German women worked full-time plucking worker bees from hive entrances to have them sting a piece of fabric. This absorbed venom, composed of a number of useful enzymes and proteins like mellitin, was then isolated and freeze-dried into a crystalline powder, writes bee researcher Eva Crane. This practice still exists today, but has been upgraded nicely by instead placing an electronic frame* in front of hives, shocking returning workers that then release pheromones to attack a cleverly placed membrane, amassing venom.

Meanwhile, during World War II, in the United States, sugar was rationed so the troops could have some. To make up for it, the number of bee colonies, both hobby† and commercial, went up to 5.8 million by 1946. Due to the surplus, commercial beekeepers diver-

---

* A similar hive gate collects pollen from bees, knocking it off their legs and bodies as they push through a narrow door. Bee pollen's worldwide popularity grew so much that western Australia will annually produce, at times, 143 tons. Global production of the aforementioned royal jelly, sought by the cosmetic industry, can reach as much as 660 tons.

† And if concerns of bee hypersensitivity prevent you from pursuing apiculture, rest assured only 0.35–0.40 percent of Americans are allergic, according to a US survey.

sified by extracting more byproducts like venom and royal jelly. And then the 1960s saw the growth of migratory services, upgrading to forklifts and semi-trucks to haul millions of bees to pollinate various crops across the nation. To help improve beekeeping in South America, African queens were bred with European bees with frightening consequences. Media scares circulated as aggressive "Africanized" bees moved north through Mexico to the southern United States, killing many people since their hybridization in Brazil.

Building a more resourceful bee via hybridization continues today. "Bee diseases have plagued the beekeeper since pre-Biblical times," wrote Eva Crane in the *History of Entomology*, "and have been partially responsible in some areas for the fact that beekeeping has not developed into an industry."

Our management of *A. mellifera* has been advantageous. But it also reveals a bit of human folly—an urge to tamper with nature without considering the costs. Fortunately, scientists, environmentalists, and individuals emotionally moved by the wonder of this creature seek ways for it to thrive today.

Buckfast Abbey lies near the southern tip of England on the edge of Dartmoor National Park, a densely wooded forest out of which a centaur or Hansel and Gretel might emerge. Getting here takes a 45-minute ride on a double-decker bus tightly hugging stone walls in the local market town of Newton Abbot and several hillside switchbacks. Trees on either side of the winding road intertwine their branches in tunnel formation, finally opening to a Benedictine church and monastery.

A breeze of linden drapes over the abbey's neatly trimmed landscape. Established in 1018 CE, the abbey was rebuilt from ruin by 30 French monks in 1882. Today the monastic grounds, replete with a restaurant, lavender garden, shops, and exhibition, receive over 300,000 visitors a year. My theory for Buckfast Abbey's popularity rests on the monks' tonic wine. The bartender at a nearby inn tells me it's like "angel's piss," as innocent in appearance as a bottle of Manischewitz, but an inspirer of "hooliganism." So mighty is the tonic wine's alcohol content that Scottish politicians have demanded its banishment.

There is, of course, an equally compelling reason for my overnight visit: the Buckfast bee.

In 1915, a mysterious outbreak called the Isle of Wight disease wiped out about 90 percent of England's hives. The native British black bee suffered the most casualties during the epidemic, which was caused by tracheal mites that entered the bees' airways, suffocating them. However, bees at Buckfast Abbey were resistant. It turns out their breed was a mix of black bee and a nonnative *A. mellifera* subspecies known as the Italian honeybee. Seeing the devastation, a then-17-year-old Carl Kehrle, known as Brother Adam, decided to breed disease-resistant, docile bees that could survive Dartmoor's hard winters.

Brother Adam had come to the monastery at age 11. The church was still in the middle of reconstruction, and after Brother Adam fell from a scaffold, the abbot assigned him to their beekeeping department. He learned apiculture from a monk who also made hive-shaped cakes of honey ginger bread. After the Isle of Wight epidemic, Brother Adam tampered with genetics, opening an isolation station near the abbey in 1925 where Italian and English black bees could mate, free from other subspecies abuzz nearby. It kept the disease-resistant strain pure and the honey bountiful. One colony alone was said to have produced 400 pounds of honey in a year, according to the 1986 documentary *The Monk and the Honey Bee*. During that time, local thieves rustled several hives. Adam told police that the missing bees, if seen, were "three-quarters of an inch in length, with dark brown and dark gray stripes." The Buckfast bees grew extremely popular. The abbey would at times have at least 500 colonies, and beekeepers favored the bee because of its lower rate of swarming from the hive. The innate sense of mystery monks possess also helped, which is why it may surprise some to learn that this once-famous, disease-resistant bee is rarely found today. The Buckfast bee is no longer at Buckfast. In fact, their Bee Department stopped breeding them, making a shift toward education in 2010.

I came here for an explanation.

From the Gospel of Maurice: "For beyond all the desires of this strange god who has taken possession of them, who is too vast to be seen and too alien to be understood, their eyes see further than the eyes of the god himself; and their one thought is the accomplishment, with untiring sacrifice, of the mysterious duty of their race."

I drop my black duffel bag on the cobblestones in front of the grand, hardwood, double-door entrance of the monastery. Minutes later, one side opens to reveal Brother Daniel, an accommodating monk with peppery white hair and a voice temporarily handicapped by a virus all the monks have come down with. (This brings to mind sick bees in a hive.) Brother Daniel worked with Brother Adam until he died. Like all monks at Buckfast Abbey, he's extremely kind, offering to carry my bag, which I respectfully refuse.

The monastery has a very distinct smell—what I can only call holy musk. A bit maple and churchy. The stone staircase spirals and turns. Our voices echo among the religious paintings, decorative tiles, and stained glass window of Jesus on the cross. It's an overcast day, and I ask how the bees are doing. Brother Daniel isn't too hopeful about how active they'll be. The weather makes it difficult to predict. "It's those nights where you can't sleep with the bed sheets on," he says of the humid evenings. After hot weather, "are the days when you get high honey crops. Our bees used to be able to get 17 pounds of nectar a day." Then he leaves me to myself.

Thirty minutes before dinner, I walk down a trapezoidal hallway. Paintings of nuns line the walls. Out the window I can see the dragon-scale roof shingles aged with rust and lichen and beyond that the high clerestory church windows. Since Brother Daniel left, I've felt a great sense of quiet and curiosity. There's a door at the end of the long hall. Monastic hums ring high, muffled in the distance like smothered power lines carrying holy energy. Vespers before supper. I'm not quite sure if I'm permitted here,* but I open the door to one of their libraries.

The silence weighs heavily as I leaf through the library's index booklet, which creaks when I open it. I find nothing on Brother Adam or the Buckfast bee. Rummaging through the books, I come across a history of the abbey from 1018 CE to 1968. There is no direct mention of Brother Adam according to the chapters, though his name is in the index. Finally, though, I discover a 1946 article, "The Culture of the Honey Bee" by Brother Adam. I get a sense of just how much the bees meant to this young boy who traveled from Germany. Halfway through, he emphasizes one of the most important aspects of apiculture: "The essential point of all bee-keeping," he writes, "is to ensure a supply of first-rate queens."

This leaves me with the lingering question: what happened?

Chimes in the bell tower indicate it's 7:00 p.m., so I hurry to the refectory on the lower level. The archways and ceilings curve. Long, linen-covered tables line the hall in a U shape. The abbot sits at the head of the table among the brothers in their black frocks. Brother Daniel nods to me as I sit by my place card away from the others at my own table. Dinner—a delicious minestrone soup—is eaten in silence to facilitate reflection. A dark ale is served, and I smear my bread with the honey specially harvested for the monks. It's sweet and bright.

Later I go into the monastery's guest reading lounge to work on some notes, and Brother Daniel is there reading the newspaper. I poke around the books on the shelf near the fireplace and see a gilded folio edition of Maurice Maeterlinck's *The Life of the Bee*. I turn to Daniel, and we talk about his time working alongside Brother Adam.

"With Brother Adam, he always had you as a smoker," Daniel

---

*I am.

remembers. "If you didn't pester him with questions, he would volunteer information as he was working. And that's the best way to learn." Daniel looks down for a moment. "He was a nice chap, actually."

"Is there a prayer for the bees?" I ask.

None that he can think of, he says, his voice hoarse. But he does mention a poem by Victoria Sackville-West he particularly likes called "Bee-Master." She speaks of bees, these *"captious folk,"* who leave their hive in a swarm. *"Follow,"* she writes, *"for if beyond your sight they stray / Your bees are lost, and you must take your way . . ."*

The following afternoon, we drive half a mile away to the mini-warehouse Bee Department, an education center with a corrugated metal roof. Stacks of empty antique shipping crates advertising "BUCKFAST CLOVER HONEY" sit atop the shelves. Next to the door is a workman's rack of apiculture tools and miniature honeycomb knives. A small group of locals are here for today's educational session. Overseeing it all is Clare Densley, a youthful middle-aged woman wearing Crocs and loose pajama pants that look cut from a hippie's Persian rug. With her stylishly graying hair cropped short, she projects a warm yet firm disposition.

Brother Daniel sniffs the burnt hay tarred inside the collection of beekeeper smoker canisters on a nearby table. "Once a beekeeper, always a beekeeper," he says. "Even now the smell of smoke— it's a nice beekeeping aroma."

Clare turns to me. "I've got a bit of a beginner's class this morning," she tells me. "Do you mind sitting in?"

Daniel mentions how the overcast weather won't be optimal for viewing the apiary. "They'll be a bit stingy today." I look dreamily at the cotton bee suits and cage veils hanging at the entryway.

"No, they won't enjoy us," says Clare. "They'll all be doing housework today, sorting the honey."

Brother Daniel heads back to the monastery. Before the Bee-keeping 101 class begins I ask Clare about Brother Adam and why his Buckfast bee began to disappear in the 2000s. We walk toward the table in the back, passing an observation hive, and she says, "It all started going to pieces."

The Buckfast hybrid bee was able to obtain its maker's pure design through what's called an isolated mating station—a land site filled with bees of the same breed. It also meant incestuous breeding. Brother Adam had traveled to several Mediterranean countries in search of species he could mold into ideal hybrids. And as Hannah Nordhaus mentions in her book *The Beekeeper's Lament*, he was "credited" for the breeding method, although less notable beekeepers had succeeded before him. (According to Clare, one magazine from 1903 advertises similar bees.) While he did license his bee-rearing technique, and some beekeepers in Denmark, Holland, and Germany currently maintain variations of the breed, the colonies at Buckfast Abbey fell victim to infectious disease. They experienced a *massive* American foulbrood problem—a bacteria in the guts of larvae that germinate and kill in the process. Antibiotics couldn't save the hundreds of colonies. Large holes were dug in the ground and the hives were set aflame in what must've looked like a funeral pyre.

"I chose not to breed the Buckfast bee," says Clare, "because I think a designer bee is weaker in lots of ways."

"I thought people viewed the honeybee as perfection," I say.

"It's so difficult to keep it perfect without getting inbreeding," she says. The makings of a healthy queen require genetic diversity, which is why she can mate with an average of 14 drones at times. If you don't let queens mate with a variety of suitors, says Clare, "you weaken her. You're looking at thousands of years before they can change any kind of characteristic."

The bees mean more to Clare than the honey they produce. The sentiments are shared by her friends, some of whom call them "hug bunnies." And like them, Clare talks to her bees. "I can imagine being a bee," she tells me later. "Sniffing around and being with all my sisters." In fact, each colony in Buckfast's apiary has a name: Mattie, Zoltana, Natasha, Layla, Aurora. The queen in the observation hive, connected with a plastic hose to an outside entrance and primed to swarm, is called Roseanne.

"My bees are mongrels," Clare continues, "but they're *lovely*. They may produce a little less honey, and they may have a slightly variable temper—but they're healthy." And while we'll soon see that designer bees aren't necessarily a bad thing, nature may have already done enough genetic grooming to deliver a reliable end product. "I just prefer middle-of-the-road bees that need less intervention," she says.

Clare turns her attention to the rookie beekeeper hobbyists. A sign with the 3-D word "WAX" covered in candle drips hangs above a metallic honey extractor. The naturalist-type woman about my age sitting next to me is named Ellie. She tells me she's started beekeeping recently because she's following "the Shamanic path of the pollen." Ellie whispers as Clare gets situated with today's handouts of helpful tips. "Have you ever chewed fresh wax with a bit of honey?" she asks, her voice breathy. "It's like gum."

"No," I say, "but I once chewed wax in a dessert." (Admittedly more labor-intensive than a Juicy Fruit stick.)

"Ooo," she sits back, impressed. "Swanky."

Class begins.

Clare brings up swarming—one of the most complex aspects of honeybees. Their division of labor is equally as impressive. Foraging, cell cleaning, resting, patrolling, attending the brood. These are all job positions that cycle through their life span. But swarming occurs once the hive is near max capacity with primary queens and virgin queens disbanding into new homes. No one better understands this process than entomologist Thomas Seeley, author of *Honeybee Democracy*.

In 1974, Seeley watched a scout bee pitch a "prospective" new home to the colony. "The bees demonstrate to us several principles of effective group decision-making," he writes, "and that by implementing them we can raise the reliability of decision making by human groups." He goes on to compare what happens to a "New England town meeting in which registered voters who are interested in local affairs meet in face-to-face assemblies."

It starts with the queen. Her eggs—these microscopic polished grains of rice—plop into empty cells. Semen from her nuptial flight with 12 to 14 or so beaux has been stored in her spherical spermatheca. Viewed under a microscope, it resembles a murky crystal ball. With only a fraction of the volume stored there, she can birth 150,000 bees in a summer. The internal colony temperature is in the mid-90s thanks to bees vibrating* for warmth or fanning with their wings near the entrance. About 45 pounds of stored honey keeps the bees' heat-generating metabolism going through the winter. Depending on the hive commercial keepers use, each colony can produce up to 220 pounds of honey annually. At some point, the bees' numbers grow so large that their home just won't accommodate anymore. Enter the realtors.

Senior foragers with their experience gathering nectar and pollen now move into the role of scouting. They're especially good at this because of their keen eye and astute sense of direction. Once the hive is full, the whole lot of them abandon ship.

A hollow tree, caves, and buildings† with big enough knotholes all make for viable homes. The scouts take several inspection trips. They walk around the surface and note the dimensions of the cav-

---

*A mind-blowing feat, as evidenced in a clip from the NatGeo documentary *Hornets from Hell*. A Japanese hornet invades a honeybee colony. As soon as the intruder attacks one bee, it's pounced on by a gang that begins vibrating, raising their body temperatures to 117 degrees Fahrenheit. The hornet's innards are liquefied.

† Rescuing a colony from a building is a task that urban beekeepers like Andrew Coté risk neck and limb for. Literally. Founder of the New York City Beekeepers Association, Coté harvests honey from rooftop hives across all five boroughs. And often he'll get a call to collect wild swarms found on skyscrapers. He told me how he once clung to a parapet wall on a 22-story building, dangling over the street with a 15-foot PVC pipe hose fixed to a low-suck vacuum.

ity's volume upon return flights, all while rallying bees to their cause back at the temporary home base, returning to potential buyers (i.e., the colony), and stirring intrigue. They perform a "friendly . . . dance off," for the crowd, which will decide the winner.

(In one of Seeley's experiments, he tested two types of hives: a 40-liter and a 15-liter box. The dance judges had to decide which scout had the better find. One bee performing stronger consecutive dance circuits, lasting 85 seconds compared to the other bee's dance in half that time, was convincing enough. The superb dancer found the 40-liter box.)

Bees rallying for the losing options either switch or give up entirely. Other members of the colony cast their votes, and the decision is made. Now the bees will help their queen lose weight before the flight. Workers bumping into her will vigorously shake her around like a fat man on a jiggling belt machine. Slimmed up from "increased exercised," the primary swarm is prepped to relocate.

Before a swarm flight, the bees' body temperature increases to 95 degrees. Scouts signal to other bees that it's time for their "buzz-run." *Life of the Bee* author Maurice Maeterlinck called them "winged quartermasters." Soon a fleet of 10,000 bees in the hive are in frantic launch mode like an airfield of World War I fighter planes eager to hightail. Only 5 percent of them know the location of their new home. As this organic cloud travels at about five miles per hour, the real estate agents, aka "streaker bees," steer the swarm, rapidly zipping back and forth like cattle ranchers steering a herd. Until the 2000s, the streaker bee—which clocks speeds of 20 miles per hour—was only a hypothesis, but advances in digital photography and computer simulations using point-tracking algorithms confirmed its existence.

From the Gospel of Maurice: "The bees have followed what at first was an audacious idea, based on observation, probably on experience and reasoning. And this idea might be almost declared to have been as important to the destinies of the domestic bee as was the invention of fire to the destinies of man."

Clare Densley focuses on the external, apiculture side of this phenomenon with her class. The top of one of her handout sheets reads, "Oooops! Your bees have swarmed. What to do?" Rehiving colonies involves carefully cutting the nearby branch the bees have swarmed to and emptying them into a new brood box so they may continue to thrive.

Clare dismayingly adds, "It could all go pear-shaped as well" ("not good" in British-ese) if the secondary new virgins aren't properly mated with, ending that original colony's story right there. Regardless, other virgins have branched off from the original colony by then and started to transform. "She gains in stature," says Clare, "and becomes not more *regal*—but—"

"Queenly?" one keeper inputs.

"Yes, a bit more assured. More confident about herself. Virgins just zip around everywhere because . . ." She hesitates because of the two young boys in the class. "You know, they're—"

"Horny?" Ellie, sitting beside me, offers.

"Um, yeah." Clare looks at the mothers. "They haven't been, um, fulfilled yet."

Part of today's lesson includes an interlude of constructing wood frames. Bees will later use these to build their combs. Clare hands out nails and hammers and a waxy sheet to jump-start the honeycomb-building. As the students work, I walk to the observation hive where Roseanne, the queen, has lost her pre-swarm weight. I press my nose to the foundation where an exhaust vent wafts scents of dried forest moss.

After the arts and crafts session, everyone suits up to visit the apiary. The 13 of us step out of the Bee Department and walk down a dirt trail into the Exmoor woods. The white-uniformed procession follows a muddy path through a lush tunnel of trees, the hive smoker filled with burnt wood shavings incensing us. We arrive at the apiary and choose two of the students' frames. Clare Densley touches upon how mongrels have better gut bacteria, and therefore healthier hives. As she mentioned earlier, the problem comes when you decrease genetic diversity. "And particularly in America they've got a very small gene pool. We wonder why we've got problems," she says, sounding more sincere. "It's our fault."

One time her hive was struck with chronic bee paralysis virus (i.e., hair loss, flightlessness, trembles). She fed the bees probiotics and garlic, and they recovered in two weeks. "I think it's the way forward," she says. But there are other afflictions still confounding entomologists. The problems certainly dwell within the bees' genes. On the walk back to the department, Clare and I revisit the legacy of the Buckfast bee. "[Brother] Daniel will never tell you, but Adam was a bit of a task master," she explains. "I think it's because Adam came here when he was eleven . . . Things were lonely for him," she adds tenderly. "He's very intelligent. And bees were a major distraction for him. It was his world, wasn't it? And I think he poured his whole life into the bees."

Bees are therapeutic. And monks can be just as mystical as bees: escaping from the modern world in search of tranquility.

From the Gospel of Maurice: "We touch here the hermetically sealed vases that furnish our conception of the universe."

The next class consists of veteran beekeepers discussing the nuances of queen rearing. A Land Rover parked out front has a "GIVE BEES A CHANCE" peace symbol on a tire cover. Nine older gentlemen sit around the table inside. A skylight filters sunlight from above. Were you to pop your head in, you might think this a mafia boss meeting of villagers taking orders from their apiary-outfitted queenpin. Clare Densley even has a faded mobster-like tattoo of two bees on her forearm.

I ask the owner of the Land Rover, Joe, a middle-aged Scotsman and "bit of a James Dean . . . crazy" type about his native country's outrage over Buckfast Abbey tonic wine. He confirms it. "That's why men in Scotland throw telegraph poles, wear skirts, and do some crazy dancing on swords," he jokes. "Who else would do that? That's why you drink Buckfast tonic wine. Didn't realize

I was 94, did ya? All the monks must be 140 years old, innit?" He nods his head, agreeing with himself. "Aye."

Later, as we visit another apiary, Joe brushes bees off the comb with a goose feather as we examine colonies, hands folded behind our backs. He turns to me: "David, would you like to hold a frame?" I've done plenty of trying things on this journey. But holding a piece of this inspired design,* viewing the intricacies up close, mesmerizes me completely.

The frame I hold is from the Langstroth hive, and I observe the teeming honeycomb in search of the queen. She's there, denoted with a blue dot on her back. A large and happy mongrel.

The next day, Brother Thomas, clearly benefiting from the longevity of tonic wine, offers to drive me to the train station in Newton Abbot. Actually, he used to drive Brother Adam to Heathrow airport when he went on global lecture tours. He cheerfully explains how stingy the Buckfast bees would get. My next stop, I tell him, is a train to Brighton. There, at the University of Sussex, a team of scientists at the Laboratory of Apiculture and Social Insects is doing a bit of queen-breeding of their own, but for a specific reason: their queens could stave off many of the ailments causing massive colony die-offs.

No other animal is more anthropomorphic than a bee. (Okay, maybe chimps.) Some examples include "busy as a bee," "the birds and the bees," or "that's the bee's knees!" Then, of course, you have poet Barbara Hamby's colorful parallels made to colonies—of which females are dominant—in "The Language of Bees."

---

* A hexagonal structure that inspired Frank Lloyd Wright's Hanna-Honeycomb House in Stanford, California, which utilizes space by connecting rooms at 120 degree angles. Less functional (but cool since they appeared in *Star Wars: The Force Awakens*) are the ancient beehive huts of stone on Ireland's Skellig Island, built in the sixth century by monks.

For the queen: *regina apiana, empress of the hive, czarina of nectar, maharani of the ovum, sultana of stupor, principessa of dark desire*

The hive: *octagonal golden chamber of unbearable moistness, opaque tabernacle of nectar, sugarplum of polygonal waxy walls*

And on communication: *a musical dialect, a full, humming congregation of hallelujahs and amens*

The first true scientific interpreter of bees is twentieth-century animal behaviorist Karl von Frisch. Like Clare Densley, he also spoke aloud to his bees. He was the first to interpret what they were saying, which won him a Nobel Prize in 1973 for "Decoding the Language of the Bee."

The discovery of their communication methods came nearly 30 years prior. To an outside observer, Frisch may have appeared odd. Sure, he did experiments on bees but, unlike others, he nurtured them as a parent, points out anthropologist Hugh Raffles. For example, he let them warm in his palm as he spoke to them. In 1919, he figured out their means of communication, attracting them with sugar water and painting dots on the backs of scouts. "She performed a round dance on the honeycomb, which greatly excited the marked foragers around her and caused them to fly back to the feeding place," he wrote. But in his Nobel lecture, he admitted that the dance's "most beautiful aspect had escaped" him. It's significantly more complex.

Beginning in the summer of 1944, the Austrian-born scientist questioned the exactitude of bee foragers' precision. How well could they denote distance? They make a figure eight known as a waggle dance. The center line, or waggle tail run, does the bulk of the communicating with the bee circling back to starting position. The run is important. While working the honeycomb "dance floor," as Frisch called it, surrounding bees take map direction and smell the type of flower to assess the quality, like company taste testers. The duration of the shake communicates distance, with a formula of roughly 750 meters per second. Anything 4,500 meters away could last about four seconds in what's called a tail-wagging dance, wrote Frisch. It was awe-inspiring. Another important bit of information is the angle of the feeding spot in relation to the hive and direction of the sun. For instance, if the hive entrance directly faced the sun, the figure eight would be completely vertical. If some dandelions were 35 degrees left of the hive, the round waggle pitches to the relative angle.

The bees' dances only added to the mystery of the species. But Frisch's discovery may not have happened were it not for a honeybee disease in Germany during World War II.

Under the Third Reich, professors were required to prove a full Aryan heritage. Frisch, however, was classified in 1941 as a quarter Jewish. He lost his position as the head of zoology at the University of Munich. At the same time, Germany struggled with an outbreak of *Nosema*—a spore-forming parasite that invades the bees' intestinal tracts, sometimes causing them to have dysentery—for years creating the same agricultural worry England faced with the Isle of Wight disease. Fortunately, Frisch had a friend in the Ministry of Food and Agriculture who knew how important his work on animal behavior would be. (A story that has close bearings to the beetle that saved Pierre André Latreille; see chapter 1.) So Frisch was assigned to investigate the massive colony losses and received a stay of academic expulsion until World War II ended.

His work on *Nosema* was "largely inconclusive." It had only been discovered in 1907. But his advances in understanding bee behavior have helped us to decode present-day issues. Honeybees have faced a long, uphill battle with ailments. In the first century, Pliny the Elder describes what we know as American foulbrood (AFB) today. Historians believe the first die-offs reported in America in the 1670s were caused by AFB. "Mysterious departures" from hives have been recorded in magazines from the late nineteenth century, including one called May disease, which is very similar to what's been the largest shake-up in the history of apiculture: colony collapse disorder, or CCD.

From the Gospel of Maurice: "The intelligent initiative of the insect has evidently received the sanction of natural selection, which has allowed only the most numerous and best protected tribes to survive our winters."

Pennsylvanian Dave Hackenberg was the first to report colonies' mysterious abandoment of their hives. Hackenberg was a commercial beekeeper for 40-plus years, taking hundreds of hives to agricultural sites for pollination. Some entomologists have disparagingly referred to Hackenberg as a truck driver who handles bees, as his hives had been aflicted with a number of bee diseases. But what he saw in 2006 puzzled him. Bees deserted 3,000 colonies—30 percent of his business—and left no trace. Investigators found, based on what little evidence there was, that such afflicted hives had been hit with multiple pathogens and an increased amount of a parasitic mite with the foreboding name *Varroa destructor.*

To the naked eye, varroa mites are just slightly larger than the dot of an "i." But to a honeybee, as bee expert Jerry Hayes has said at multiple honeybee-related conferences, varroa is the human equivalent of having a "parasitic rat on you, sucking your blood." These virus vectors deplete bees' immune systems, physically weakening them with afflictions such as weight loss and deformed wings. But varroa, discovered nearly a century earlier, first caught our attention in the United States in 1987, so pinning them as the sole cause for CCD, nearly two decades later, was difficult.

The harrowing mystery of CCD carried on for the next two years, spawning documentaries, reports from major networks, magazine features, and various "Beemageddon" hoopla surrounding the bees' steady decline from 5.8 million in 1946 to 3.3 million (1990) to 2.4 million by 2006, notes Hannah Nordhaus in *The Beekeeper's Lament*. CCD then hit Europe for the first time in October 2009. Since CCD, colony loss surveys have been taken biannually in the United States, given the public's dramatic awakening to honeybee health and how fragile these creatures are. The number of colonies rose to 2.6 million in 2015, despite a detrimental annual colony loss rate of 44.1 percent.

Dennis vanEngelsdorp, the University of Maryland entomologist who helped Hackenberg diagnose CCD, doesn't see many case-specific diagnoses of CCD today, making the sudden jolt all the more strange. However, mass die-offs still occur for various reasons. In a more recent survey, from 2015 and 2016, the winter loss in Europe was 11 percent. (Admittedly vanEngelsdorp is worried that Europe's survey-taking techniques differ too much for accurate comparisons.) Canada's loss was 16 percent. And the United States suffered a 28 percent loss that winter, discouraging hobby beekeepers and closing businesses that couldn't afford to replace hives. Nailing down the cause, therefore, is urgent. "I think CCD brought to light that there is a *real* problem with bee health," says vanEngelsdorp. It also revealed how "complicated" pollinator health can be altogether.

Figuring out the nuances as to *why* takes equally complex research. Jeff Pettis of the USDA Bee Research Laboratory in Beltsville, Maryland, tampers with hives with various methods to deduce what is killing bees: climate, soil moisture, foraging, and pollen protein content. A type of pesticide class known as neonicotinoids, which confuses the bees' ability to recognize flowers and crops, has long been cited as a cause of colony loss. Although neonics are usually applied as a seed treatment, footprints of the pesticide remain in the pollen and nectar. In one 2012 study in France, neonics were found to be highly toxic to bees, stirring great debate (often rallies with bee-costumed environmentalists* shouting into megaphones) about the chemical, which eventually led to an EU ban on three insecticides containing it.

But the evidence is questionable. Many scientists directly dose the bees with neonics, which hurts the environmentalists' argument against neonics.

One study validating these neonic claims came from Italy, where researchers sprayed commercial admixtures on Spanish chestnut leaves and then put the leaves into cages containing 30 bees. The 2011 paper describes how one insecticide of the neonic class called thiametoxam "caused total mortality within 6 hours" in concentrated doses. Toxins in another neonic called clothianidin "caused extensive vomiting" as well as tremors, staggering, and delirious movement. But like many similar experiments, this was done in a lab setting. One meta-analysis of 14 similar published studies on another banished neonic called imidacloprid showed that "a clear picture of the risk [neonicotinoid insecticide] poses to bees has not previously emerged." Investigations, author James Cresswell said, had an "inconsistent outcome," pointing out that the pesticide studies are done over a short time and don't focus on sublethal impacts of long-term exposure.

A broad-spectrum analysis by researchers at Pennsylvania State University found over 170 different chemicals in hives. However, as entomologists Diana Cox-Foster and Dennis van-Engelsdorp wrote in a *Scientific American* article, "healthy colonies sometimes have higher levels of some chemicals than colonies suffering from CCD." And some CCD cases shared attributes of one common affliction called the Israeli acute paralysis virus (IAPV). This virus strain, which solely attacks US bees, probably arrived from Australia in 2005. But while "the infection mimicked

---

*While some environmentalists protest rightly with well-researched arguments, there are the overly passionate and straight-up weird, e.g., those who blame it on cell phone tower radiation or claim that "bees were abducted by aliens." 'Nuff said.

some symptoms of CCD," not all exposed bees die the same way. It could mean that CCD may require a cocktail of factors* or that some bees are IAPV-resistant.

So I ask again: what solution is there for the honeybee? Many beekeepers place blame on human folly,† believing that such massive die-offs result from piss-poor management. "I don't think commercial beekeepers could be 'piss-poor beekeepers,' otherwise they would go out of business," counters vanEngelsdorp. "It's a complicated situation, which means it has a complicated solution . . . There's some mortality that's *happening* that has nothing to do with management."

Tracing this "happening" can be nerve-racking.

A 2016 study from Free University Berlin caught my eye. Researchers attached transponders to bees, monitoring them in a 900-meter radius field with a harmonic radar system. They then filled feeders with neonic thiacloprid sucrose solutions. What they found was that their untainted control bees "consumed 1.7 times more sugar solution per day than treated bees," performing nearly double the amount of foraging trips. The fact that concentrated doses were harmful is not surprising. What was curious, at least to me, was this idea of tracking bees with chips—an interesting idea that could illuminate a microuniverse.

Paulo de Souza is using this idea to work with researchers across the globe. In a small lab room at the University of Tasmania, a sedate honeybee rests on an icepack like a patient prepped for surgery. Just removed from the lab's freezer, the bee's metabolism

---

*Stressing the earlier remark of the importance of healthy queens, one 2016 study discovered that low sperm viability can encourage colony loss. Queens bred by different businesses and later shipped in tiny boxes by UPS and the USPS were found to lose over 50 percent of their stored sperm with temperatures around 40 degrees Fahrenheit and above 104 degrees.

†In a move to kill mosquitoes potentially carrying the Zika virus, millions of bees were ravaged in South Carolina with an insecticide known as Naled. According to a bee farm owner interviewed by the Associated Press, the collateral aftermath was as if her field had "been nuked."

has been drastically slowed long enough—about three minutes—for a lab assistant to delicately glue a radio-frequency identification (RFID) sensor to its thorax without the risk of being stung. Soon it will rejoin 5,000 honeybees in several hives, each tagged with a 2.5-millimeter backpack. The goal? Capture microscopic real-time information on bee ecology, and ultimately the world.

"Each honeybee has an ID tag, like a license plate," Souza says about the experiment. Funded by Australia's Commonwealth Scientific and Industrial Research Organization (CSIRO), the honeybee backpack project took flight in 2014. Some RFID bees have shown up as far away as Chernobyl. The five-milligram RFID computers have a clock and memory for storing measurements so the research team can detect when bees depart and arrive at feeder stations. But by using Monte Carlo simulation software, the researchers are able to predict the 4-D ecological information they'll obtain in the future, such as the impact of weather in regard to the bee's weight loss and decrease in numbers. This type of airborne microelectromechanical system is comparable to "smart dust." It can detect atmospheric gases, temperatures within microclimes, and magnetism. Through quantum and statistical mechanics, information via smart dust can help explain internal energy, heat capacity, and any other thermal dynamic behavior of an ecological system on a particle level. As tiny weathermen, bees may also expound on the major factors killing our planet in real time.

If successful, CSIRO's environmental monitoring may finally help us understand massive bee die-offs, since the scientists are introducing hives to pesticide-ridden lands and bloodsucking mites. (Though *Varroa destructor* has luckily not hit Australia.) These mites' first appearance in the United States coincides with the grand decline of colonies, and they carry an array of viruses with interesting names: Lake Sinai, deformed wing, Israeli acute paralysis, slow bee paralysis, Kashmir bee, black queen cell, chronic bee paralysis.

Although bees have evolved durable bodies over millions of years, human-propelled ecological interference renders them feeble. In the past, a bee colony could still survive, even after 20 out of 100 bees were infested with varroa mites. Today, you have to worry if you find 5 out of 100.

"You wouldn't let your *dog* die from pests. You shouldn't let your *hive* die from pests," says vanEngelsdorp. "If you really want to breed for resistance, and I think there's a lot of evidence that there are resistant bees out there, the responsible way to do that is to monitor your mites and breed from the ones who had the lowest population [of mites]."

One Monsanto employee is closer to finding a solution for the collapse. As reported by science writer Hannah Nordhaus, Monsanto's Beeologics lab is trying to use an RNA interference technique to modify crops that will directly target the varroa mite's genome. Jerry Hayes, who heads up Beeologics, tested this specially modified RNA in sugar syrup, and is currently doing a trial with over 1,000 bee colonies across the United States.

There may also be another solution that involves a technique used over the past 130 years: captive queen-breeding.

The Honeybee Breeding Laboratory in Baton Rouge has been producing queen bees with varroa-sensitive hygienes (VSH) since 2001. VSH-imbued worker bees can detect the destructive mites and remove them. Bee expert Patrick Heitkam in Orland, California, produces 1,000 queen bees per day. Some breeders can make up to 3,000. University of Minnesota entomologist Marla Spivak also breeds for hygienic queens, saying that we shouldn't perpetuate bees requiring chemicals and antibiotics.

It's this type of queen-rearing that took me to the English seaside city of Brighton—more specifically a renowned lab populated with minds from across the world contemplating how bees might adapt to the environmental constraints we've placed on them.

Fragments of golden sun bleed through the blur of leaves as I prop my head upward, elbow resting on the train's armchair, riding into

East Sussex. Rapidly passing dairy cows and riverboats, I approach the quaint hills of green where Winnie-the-honey-bowl-headed-Pooh was created. As we pull into the Brighton railway station, blue beams lift the glass pitched roof high above, braced by a web of steel.

At the bus stop, I meet a woman who's also heard of Buckfast Abbey's tonic wine. (Man, this stuff gets around.) I tell her I'm here to have a quick lunch with Professor Francis Ratnieks at the University of Sussex.

A few hours later, up two flights of stairs, I'm writing in a sloped-ceiling bedroom. A bee lands on the inside windowpane in front of my desk, crawling across the blemished lens of terraced homes. It buzzes frantically against the base, attracted to the sun rearing from the clouds. In the past, I might've tripped over my chair in a frazzled panic of getting stung. But this time, I jail the bee in an empty teacup, slide my notebook cover beneath it, and release it back out the window.

The next day, my bus deposits me at the university stop for my noontime lunch at Laboratory of Apiculture and Social Insects (LASI). I slosh about with my grocery bag of chicken salad and crisps ("potato chips" in American-ese), but I can't find Ratnieks's building. Or really, I have, but mistake the bland brick building several times for the electric generator room tucked in the campus's corner pocket—which it kind of is. There's near to zero signage except for a crud-covered post on a side gate that reads like a Neighborhood Watch sign. It says "DANGER: STINGING INSECTS," and depicts a silhouette of a bee with a hemorrhoidal lightning bolt.

This must be the place.

My pounding on the white doors is muffled by the loud machinery beyond them. The doorbell has been removed. Fortunately, two students carrying lunch come up and keycard me in. LASI has had workers from across the world: Syria (that'd be Hasan, whom I meet in passing, whose full-time job is rearing their special hygienic queens), Germany, the United States, Italy, Russia, and France.

I mosey through the tight-knit LASI workspace. Cabinet shelves next to Ratnieks's office act as a sort of Bee Shrine with various souvenirs and tchotchkes: Winnie-the-Pooh sippy cups, a liqueur mead growler, honeys (black, Spanish, Danish, lavender), 13 bottles of mead, a shiny bee tie bar, emptied honeycombs, and plush bee dolls. Outside in the slender garden on the opposite side of the gate's warning sign is a flower patch of 14 friendly varieties, including the Velcro-y bulbous *Helenium* and purple cobs of catmint. All 14 varieties are part of LASI's collaboration with nearby parks to promote garden flowers that are "100 times" better in helping insects.

Francis Ratnieks finishes writing an e-mail and escorts me back inside to the queen-rearing room. "This is a new venture," says Francis. He stands over six feet with a mellow voice that carries as though his mood ring were a permanent dark hue, touting bits of bee knowledge as a sort of sage. Francis even visited the same stingless bee apiary as me in Brazil. As a demonstration of his prowess, he points to a bee forager performing a waggle dance in an observation hive. It repeatedly cycles in a figure eight pattern, and Francis instantaneously analyzes it: "That bee is communicating a distance of about 400 meters at 80 degrees left to the sun," he says. "It's probably bramble flowers."

The room is full of small nucleus hives, intended for queen-mating. Like humans, bees learn to associate. They can pair colors and odors with food, navigate toward specific patches of flowers from the hive, and identify landmarks near the colony's entrance. But hygienic behavior is innate, so finding hygienic bees takes meticulous vetting. "Hygienic bees remove dead and diseased brood from sealed cells," Francis says, "and this confers disease resistance. And yet this behavior is quite rare and highly variable." Five to 10 percent of the colonies' honeybees are hygienic. Experts at LASI find them by making a frozen patch of dead brood. Hygienic colonies will quickly remove them. "You could literally breed for that trait," he says.

So far they've shipped 80 hygienic queens and have orders for 80 more after only a month. Francis prefers using a nucleus hive over artificially inseminating queens—an intensive practice requiring intricate tools and the steady hand of a watchmaker. But artificially inseminated queens don't live as long, making it harder to maintain your stock. However, the folks at LASI intend to use instrumental insemination to rear hygienic virgin daughters and drones.

We walk back outside to drink our coffee while sitting on a bench across from the Bee Shrine. "So," I start, "do we depend on these bees we've domesticated as much as they depend on us?"

Francis immediately corrects me on my use of the word "domesticated." In French, bees may be called *l'abeille domestique*, but that's a far cry from any farm animal or dog. "[Bees] have been highly modified from their ancestors," he says, "but I think honeybees are hardly different to the way nature made them. We have rather little affected honeybees. Almost everything that they do is part of their natural behavior."

This is why hygienic behavior in bees, limited in the United States as they are for now, is so exciting. Had beekeepers not controlled diseases, and just let them be, sure, there would be more colony die-offs, but eventually evolution would favor hygienic bees. "The United States," says Francis, "is a place that believes in chemicals" and antibiotics to manage various diseases and varroa mites. "But in the long run, they will not be successful because of resistance. Antibiotics tend to mask the disease rather than fully cure it."

I am surprised to learn later that Francis doesn't share the sentimentality of his bee behaviorist colleagues like Karl von Frisch. I ask him: "What fascinates you most about them?"

"Well, you're talking to someone who's obviously biased," he says. "There's so many angles to them . . . They do such incredible things. For instance, the other day I saw them police each other to resolve conflicts over reproduction. The honeybee has the most sophisticated communication signal of any organism other than humans. It also has a very strong link to humans." Francis then points to the wall and recites a number of bee quotes hung on it. Charles Darwin. Yeats. The Qu'ran. My bee wordsmith Maurice Maeterlinck. And of course Frisch, who called them "a magic well for discoveries."

What Francis says next sums up my general feeling toward all bugs.

"People think interesting species exist but not usually where they live or in their national park," he says. "Whereas, wherever you're living—if that's North America or Europe, Australia, Africa, South America—you can go outside and you can see honeybees foraging within yards of where you're living. And I would say that that common animal is the world's most interesting species. Maybe second to humans." He looks down at his mug of coffee. "Um, of course I'm very biased"—he pushes his glasses up the bridge of his nose—"but there you go."

From the Gospel of Maurice: "This secret spring comes from the beautiful honey, itself but a ray of heat transformed, that returns now to its first condition. It circulates in the hive like generous blood. The bees at the full cells present it to their neighbors, who pass it on in their turn. Thus it goes from hand to hand and from mouth to mouth, till it attain[s] the extremity of the group in whose thousands of hearts one destiny, one thought, is scattered and united."

Francis and I join the lab students at a large table filled with sack lunches, brownies, and marmalade cake. Chatter evolves into a discussion about neonicotinoids. Banning neonics, as the EU has, could result in a backlash where farmers resort to old, more harmful pesticides like DDT and other lethal chemicals. The same ones used 60 years ago when 82,000 colonies died in California in one year. One of the LASI students, Nick Balfour, recently wrote his thesis paper about a long field trial with neonics. He found the insecticide did little to affect the colonies.

"When I see the way commercial beekeepers keep their bees in the States—move them thousands of miles—I find it hard to believe that's not harmful," says Norman Carreck, Francis's colleague and a beekeeper since age 15. "A lot of losses ought to be preventable. There are a half dozen commercial beekeepers in the UK who've got a couple of thousand hives. Most of the others have a few hundred. We have nothing on the scale of these big US operations," he says. "The British ones move bees but not very far."

The theory directs attention to two overarching factors that can escape us: nutrition and stress. Dennis vanEngelsdorp suggested a simple solution—not a magic bullet—that the folks at LASI also promote, and it's one you and I could do: start a pollinator garden with insect-friendly plants that encourage biodiversity—even if it's small. Insects are married to specific plants. Research local ones in your area, and then lay down a diverse garden. In 10 to 20 years, it could make a significant difference. When I asked vanEngelsdorp if it was tough being an entomologist and if he was *optimistic* about the bee's future, he outright said, "Oh, yeah. Oh, yeah. If I wasn't, I couldn't do it."

On my journey I've visited numerous labs and rearing rooms, bug artisans and maggotologists, comedic exterminators and beetle smugglers, cricket chefs and mosquito launchers. But the one missing facet has been a taste of the Old World.

Following Carreck's line of thought, I make my way to London for a flight bound for Greece.

Before leaving the UK, I stop at Kew Gardens—a 300-acre, botanical behemoth—to see a 55-foot sculpture called the Hive. Created in 2015 by Wolfgang Buttress, the 170,000 aluminum poles spiral into what the artist has called a skeletal "latticework," creating an interior dome lighted with hundreds of LEDs and a glass floor with hexagon patterns. Inside is an oxytocin-fueled sensation. The lights' intensity and the low pipe organ drone vibrating through the aluminum pieces is actually an interpretation of sound and activity transmitted via an accelerometer placed in one of Kew's beehives. As the sun peers from the clouds, the bee-generated music swells and grows, resonating in this cathedral. Hundreds of attendants exit and return to the Hive, eyes open and necks craned in reverence.

Epilogue

At 8:00 a.m., the Mediterranean Sea is a rolling mass of black marble. It would be an opaque marble were it not for the twinkling glimmers of sun and the water-churning propellers of the *Nissos Mykonos*, a ferry docked in the Port of Piraeus. The foghorn sounds its departure. I board the ferry via the aft dock and watch the apartments and villas layered in sun shades become smaller. The propeller-generated trail of sea froth fades from the mainland. And the horizon where Athens remains is a thumb smear of olive oil murkiness.

So why am I sailing to a 99-square-mile island called Ikaria? A while back I read about Blue Zones—one of the many health crazes from the 2000s promoting Mediterranean diets. What made Blue Zones stand out is that researchers found places across the world whose inhabitants happened to have greater longevity. Ikarians held a place in their heart for a special honey made from the nectar of heather flowers called *reiki* or, as the locals refer to it, *anamatomelo*, aka anama. Thick as peanut butter, it reportedly boosts the preternatural longevity of Ikaria's inhabitants. This once-unmarketable bee by-product that only locals would take as a daily "vitamin" or medicine is now a key attraction for health nuts touring Greek islands. Other honeys on the island, not necessarily as mystic or purportedly life-extending, are pollen and nectar cocktails of heather, pine, fir, strawberry, orange, wild lavender, and more.

Doing some more digging, I found a blog post about Lina Tsingerlioti, a young, city-raised woman attracted to Ikaria's simplicity. The article mentions her recent beekeeping hobby. After a couple of false leads, I was able to track her down. I expressed my interest in visiting the island and tasting its fabled honey. Greek honey has stood out since ancient times as the most "prized" and "distinctive," writes *Sweetness & Light* author Hattie Ellis. *Thymomelo* is one such remarkable honey nurtured from thyme. (Actually, honey is regurgitated from the mouths of bees. But "nurtured" sounds prettier.) Lina and I corresponded for several months before my trip to Europe, and she gave me the addresses of two of the island's best beekeepers: Giorgos Stenos and Yannis Kochilas.

After weeks of planning, I find myself on the *Nissos Mykonos* on the way to meet the beekeepers. The Aegean is vast and its shorelines distant. The ferry calmly cuts through the liquid sapphire sea, its engines emitting a mechanical rumble like a cross-legged robot chanting *Om*. We stop at two islands: Syros and Mykonos—tourist destinations with bleached-white, LEGO-brick homes snapped into the mountainside. When the *Nissos Mykonos* reaches Ikaria's Evdilos Bay, the travelers gather in the ferry's garage, luggage in hand, and watch with anticipation as the 20-foot cargo door lowers in the port. The ardent sunlight tightens my skin as I round the cove. Small cafes and confectioners and shops line the streets with locals sitting at tables langorously conducting the air with open hands mid chit-chat. The island is entirely mountainous with serpentine gravel roads. Although many townsfolk hitchhike to get around, I've elected to rent a car—more precisely, a supercompact Chevy.

Exotic Ottoman classics play on the radio as I weave around the coast toward my hotel. The mountain flora is parched. The spines of juniper trees elegantly twist from the ground, their branches reaching out. I learn from a local beekeeper named Xenia Regina that the island has a plague of wild goats, or *rhaska*, some of which I see on the road's edge as if they are hitchhiking. (Apparently government incentives for owning a certain number of goats has led to their overpopulation.) There's also been a concern over the growing number of tourists. "Thousands of people coming here to learn the secret of longevity," Xenia told me. "It's not that simple. You come here, they tell you some of the secrets, you buy some honey, oil, and wine and herbs—and that's it, you'll be 100 years old," she says.

She attributes higher life expectancy to sociality. "[Ikarians] go to all the festivals, have gardens, watch grandchildren, dance at soirees," she said. "It's also like the laws of nature. The stronger ones survive." (Interestingly enough, Giorgos Stenos taught Xenia apiculture. In fact, he's taught Lina Tsingerlioti, my electronic pen pal, and anyone on the island willing to learn, imparting his secrets for harvesting such quality honey that only a select few have been fortunate to sample. His assistant was quick to state he was "the guru of beekeeping.")

I reach my destination, Atsachas, a hotel and restaurant along the cliffs of Livadi beach. Its water clarity baffles me. I can only relate the crisp aqua palette to the fabric paint section of craft stores my mom would drag me through. For a moment I ponder how much dye would be necessary to achieve this beauty. Eugenia, the owner of Atsachas, greets me in the kitchen. She refers to me as "Colorado."

"The rest," she says, waving it off with her hand—the rest being my name, David MacNeal—"is too difficult to remember." Later I become visibly panicky trying to figure out how I'll coordinate meeting Giorgos since my Ikaria connection, Lina, has briefly stopped e-mailing. This edgy feeling comes on, but Eugenia is quick to offer that Giorgos Stenos is actually her cousin. Meanwhile, her 20-year-old son Teo rolls a cigarette and offers to make me an iced coffee. He tells me how he intends to move to Athens to learn apiculture—the apple not falling too far from the tree.

After settling in my room, I swim in the sea, buoyant from the salinity levels. As I lay out on a lounge chair, the constant crash of the waves and sunshine lulls me. I pass out for what feels like an hour . . . but is actually three.

The next morning, I sit outside my room smearing store-bought Ikarian honey onto bread. Coffee brews in a briki pot on an electric hotplate. The beach is a flight of stairs away, in plain view. Directly in front of me I watch as bees forage from the succulents sparsely flowering in the garden plot decorated with white rocks and arranged in a spiral. This playfully bright honey is a blend of different plants and seasonal harvests. It's fragrant and delicious. As of that moment, it's the best I've ever had.

Bizarrely enough, when I meet up with Lina, we come across the other beekeeper, Yannis Kochilas. He is revving up his motorcycle in Christos Raches square, a rock-tiled enclosure in a mountain town 2,400 feet above sea level full of small tables, street-lamps, feral cats, cafes, and rustic villas. Lina translates the conversation in Greek for me. Turns out Yannis was about to write back to me, but then the Ikarian time paradox got to him—the one where one hour turns into three. And he forgot. That's how, before sunset, Lina and I find ourselves in Yannis's foliated home in the mountains.

The narrow street intended for donkey carts and traveling merchants is quiet. Cows bellow in the distance. A rooster crows. And a choir of crickets and cicadas sing. Flowering trees consume Yannis Kochilas's home. Slabs of Sheetrock are slated on a roof on flagstone walls. The ancient honeybee clay pot Lina and I stare at is similarly enclosed. It's called a *hastri*. "You can see on the mountain still, a few left," she says of the hive from bygone eras. "But the people, sometimes they steal the clay pots. They make them chimneys for the houses."

Yannis, a 53-year-old man with thick glasses and a wing-shaped T-shirt sweat stain, steps into a shed by the hastri. He comes back from his "honey house" holding Old World beekeeping tools. One resembles a dustpan-sized hay pitch. This wooden tool was used to hold the combs. The L-shaped metal crowbar would cut and separate the hexagonal honey stores from the combs containing the incubating brood. These tools go back in his family four generations, back when they had 10 hives at most. Today he keeps 150 hives.

He invites Lina and me into the honey room. Similar tools are meticulously displayed on the walls as though the room is a mini-museum: An old swarm cage, easily mistaken for a fencing mask; a rusted honey ladle; and a hive smoker that looks like a perforated kiln, once fueled with dry bull dung. Displayed alongside his collection are bee photographs as well as honey competition awards from Italy's BiolMiel annual contest.

"Four samples I sent, and four I was awarded," says Yannis. Unlike Giorgos, Yannis is more private in his beekeeping affairs. "The older beekeepers look at you like a competitor." He just started entering his honey in competitions. His thyme honey took first place in 2014 and 2015. Yannis harvests little, but the jars of honey are worth the price. He charges 20 euros for his thymomelo, which is available only within Greece to select subscribers. His process involves waiting until the sea is calm, sailing out to an island, and climbing cliffs with a hive backpack that he'll lay down in a flat area. He's made these trips for the past 15 years.

"It's more for his satisfaction that he does it, because it's too hard," says Lina. "He likes the situations of going to some rogue islands. It adds a good quality to the honey." Other beekeepers new to Ikaria are intimidated by the footwork required as well as the aggressive behavior of the bees. But Yannis is devoted to the hunt.

"When you attend the hives, you forget everything," he says via Lina. "You don't have contact with the rest."

Yannis returns from his house carrying spoons and plates. Jars full of melted auburn, topaz, and sun-colored crystals stand on the table. The first organic honey we try is heather and strawberry. It's thick and light with hints of vanilla. "It's kind of rare to find pure heather honey," says Lina. "It's an honor, you might say, to try this."

The next jar is pine tree and wildflowers. "Wow!" I blurt. "I—I feel this tingling on my tongue."

Characteristically, what separates Greek honey from the others is the addition of pine. "It's by the extracts of an insect that lives on the pine trees," Lina clarifies. "And this insect lives only in Greece and Turkey." Bees forage from the similar extract that comprises shellac. The scale insect involved (*Marchalina hellenica*) produces a sweet substance from pine trees that is found in 60 percent of the honey from those two countries.

The last one we try is 100 percent thyme. It's aromatic. You can instantly smell the herb. "For this he transfers hives to little uninhabited islands." And when I taste it, I go gaga. My eyes widen. Such a rarity is this taste that not even UC Davis's "Flavor Wheel"* lists thyme as a foraging source. So here's what thoughts run through my head: *Tilt-A-Whirl carnival lights. A spring fountain. Sap-coated balloons. Bee blood. Leaves. Herbs, herbs, herbs.*

As I'm clearly unable to put the deliciousness into words, Yannis leaves me with two jars: one thyme and the much-fabled anama, aka reiki—a mana force of heather as thick as a giant's earwax and as tasty as nature's finest candy. Before Lina and I depart, we sit in his courtyard drinking a strong, clear pomace brandy derived

---

*Incredibly accurate, the wheel breaks down 100 descriptions, sommelier style. It starts with primary reactions, i.e., fruity, spicy, floral, animal, herbaceous, and branches into specifics, ranging from practical sources (jasmine, fig, clove) to unorthodox (cat pee, humus, locker room, and goat).

from wine grape extracts called *tsipouro*. Mosquitoes keep attacking me, and the two laugh at my reactions.

"You have sweet blood, we say," Lina smiles.

I take another sip, gesturing toward Yannis. "It's all the honey
I've been eating."

When Lina and I drive back to the town square in Christos
Raches, we pass by a number of locals she knows (as well as a man
walking a goat). We stop and she gets out of the car and talks with
them. The communal friendliness is expected in an island population of 10,000 or so—there isn't much room for enemies. The in-
timacy is furthered by the large number of festivals, like the
summer solstice festival in June. Flower wreaths will be burned,
and as custom dictates, men and women will jump over the fire
pit, which saves them from bad omens, as *bouzouki* guitars are
rapidly strummed and townsfolk link arm and arm in a growing
concentric spiral of dance.

In the afternoon, my friendly guide and translator Lina Tsingerlioti and I hike up a hill of switchbacks to meet our other famed
beekeeper, Giorgos Stenos.

We reach the gate of his base of operations—a rundown building that aspired to be an auto shop—and enter. Inside, the *what-
do-you-need* accoutrement of products ranges from rice satchels
and fax machines to screws and bolts or dish soap and binders.
Half of the store is organized, but the other half contains clutter
from what looks like an episode of *Hoarders*. Lina calls out for the
guru, her voice bouncing off the concrete walls, muddied tile, and
paint chips. He appears from a back room. I shake his hand. His
skin is a tan leather satchel, sun-creased and tough.

This is Giorgos Stenos. At 84, his laugh comes through like a
crackly wheeze over radio static. He stands with the vigor of a teen-
ager. I thank the Ikarian master for taking time to speak with me.

"No," he says, via Lina, "I've had bees for 66 years, and for that
reason I know a lot of things. I like to transfer my knowledge to
people since not many want to do this." When I ask how many

hives he keeps, he says that like the French, it is bad luck to say. But the ballpark figure is 100. "How old are you?"

"Thirty," I tell him.

"Triánta," says Lina. He has a granddaughter a little older than me, he says, noting I'm old enough to be his grandson. Giorgos took up beekeeping in the 1940s after reading an article. "I decided then that I wanted to deal with these animals because all the others do harm somehow. But the bees only do good." Nurturing became second nature to him, especially as Ikaria suffered during World War II. "There was no food," he says. "The island became isolated because of the invaders taking the boats. Even if you had a bag of gold, you couldn't find lentils." At age 10 he took care of his siblings. "People had to cultivate to survive."

He was the first beekeeper in his family. By now he's trained over 50 pupils, although he's still learning himself. "After 60 years, I keep having surprises," he says. "Because I'm self-educated, I had to be very observant. And no matter how many years you are beekeeping, we will never really understand the way they behave and their instincts. The bees were here for millions of years before humans. They survived without my help or your help," he says, pointing, "or her help. We can try to understand what they want, but we could never really control them." He sits back in a chair at his desk, a rugged metal beast you'd see at NASA circa 1960. "Love bees as they were your mother or child or yourself. If you love something, you take care of it . . . What [humans] can offer is to move bees to hot places with food."

Before Giorgos can give me a jar of honey, he says there will be a test. "Not to anybody," he says, wagging his finger about it. "First I have to examine the person. I have special equipment from France," he says, ribbing me.

"And what is he looking for?" I ask.

"No, no," he holds up a hand. "Only the doctor says, and it's a secret." Like me, he laughs at his own jokes. "One of the basic things I'm looking at is how much the honey is boiling your blood," he continues. "To see how much you need. Because if you take too much, you might have problems." Needless to say, I'm enamored of my boisterous host. He directs Lina and me to a room with two 40-kilogram vats out of which he pours fat globules of honey into jars. We bring them back to his desk with plastic spoons. Lina grabs paper towels and wipes part of the clutter off his desk: spilled honey, stained appointment books, and an ash-filled plastic cup and papers dog-piled over office supplies. We taste the honey: *Granulized sugar. Molasses. Heather flowers. Vanilla soup topped with maple syrup.*

"I'm kind of falling in love with this one," I tell him. The glint grows brighter in his eyes as he slaps his thigh, laughing. I ask the guru about who he plans to pass the torch to. Embers of that fire, his "wealth" he says, have been given to the pupils all this time. "He says you can come after 30 years to discuss it with him," Lina says, offering me the job position.

"And you have passed this little examination," she continues. "Giorgos," she says, explaining the old man before me, "feels honored that his parents gave him this power to love people." In a way this "life sap" is also an expression of love. "You have to be careful, now, with the honey I gave you," Giorgos addresses me via Lina. "How you eat it, with whom, and what girl you eat it with."

With that we shake hands, and Lina takes a photo of Giorgos and me. Occasionally I'll pause on it when scrolling through my iPhone. Giorgos reminds me of my grandfather, a man named Claude Surur, who before passing away fueled the flames of my grade-school curiosity. Something that's never left me.

It's true the world can be lonesome. But sometimes you lift your head, or travel halfway across the world, and there's somebody there. Sometimes you talk to them. Sometimes you just share a moment. And the moment is well worth all of it.

As Lina calls her dog, Cinnamon, Giorgos walks me outside his general store to a small patch of mountain flat where his hives are stirring. Blue paint peels off the individual boxes. He points to the bees, the exchange between the two of us nonverbal. At the hive entrance some workers edge their way in and out of the colony. And Giorgos Stenos looks on admiringly, as though watching his daughters depart to achieve great things.

When I return to Atsachas, I'm greeted by Eugenia in our established rapport. "Colorado," she says, "what would you like to eat?" As she makes the "Greek lasagna," *pastitsio,* I sit in the courtyard. Her son Teo, the future beekeeper, welcomes me at the table—"Yasou, David"—bringing an empty glass for the strong Ikarian wine Giorgos gave me as a parting gift. His own vintage, kept in a plastic olive oil bottle. The glass is small, so I compromise with many pours that come out in a translucent red. The evening dark silences the cicadas' electric hums as the stars flicker above, leaving only the chatter of sporadic conversations from surrounding diners and the ebb and flow and break of the seawater against the cliff and beach below.

I think about this change stalking me since the outset of this journey. This realization about our relationship with these micro lever-pullers of the planet. They are the imperative hem in our life fabric. For this reason, our relationship demands pause and observance. For those who allow this break in the barrier, the tumbling of some wall, you can see what I see. Just stop midhike and watch and wait. You could see them at work. Crafting. And in a way I find it oddly comforting.

So I drink the wine, smoke Backwood cigars, and smear fresh tzatziki on bread. The air is thick, the humidity clinging, and the bugs are out. Hard-shelled ticks drop into my curly hair from the thatched-roof patio; crickets sing; beetles buzz across the ocean cruise line lights hovering in the distance; ants and cockroaches race at my feet; flies perch on my plate; flapping moths and suspended spiders obscure my vision; and the endless high-pitched squeals of mosquitoes kamikaze my ears. The bugs around me celebrate. I am bitten again and again and again. And I drink this powerful cup.

Acknowledgments

Writing a book can be impressively isolating. Were it not for my family and friends, it would be an impossible journey to undertake. (Sorry, Bill "Fucking" Murray, but even cockroach companionships have their limitations.) I'm about to go into a lot of thank-yous here, but perhaps the largest I owe is to my mom and dad, who supported my journalistic pursuits despite my times of self-doubt. And then there's my sister Kristen. How much your impromptu flight to Denver to lift my spirits meant to me, you'll never know.

Bugged happened because of you guys. This especially goes for Tony Bellah. Our near-daily phone conversations, averaging at about 40 minutes per, were motivational sparks fueling the creative fire of a book about bug people from conception to proposal and through the completion of the manuscript. A large part of this book is also owed to Reilly Capps—my fellow journo buddy and Couchsurfing host for when I first moved to Denver. Your ideas, suggestions, and encouragement were not only influential but gave me faith in pursuing this project.

To that end, I'd like to thank my agent, Eric Lupfer, and incredible editor, Elisabeth Dyssegaard, who, like me, appreciate weird things. Both of you took a chance on me. (Shout-out to Laura Apperson, Bill Warhop, and Alan Bradshaw, as well, for aiding in book production!) Thank you for believing something as niche as insects could appeal to a wide audience. Because insects are truly amazing in every definition of the word.

Drinking the buggy Kool-Aid along with me was Michael Kennedy, who hammered out a series of masterful illustrations. Each one surprised me like a bunch of Christmases wrapped into one. I owe a great debt of gratitude. In fact, the whole Kennedy clan is beyond lovely, so big thanks to your wife, Emily, and daughter, Charlotte, for just being a cute baby with a baby face. Handing over the mic to Mr. Michael Kennedy for a second, he'd also like to thank his wife, Emily, for her endless support and enthusiasm.

Then, of course, there are the entomologists and bug-lovers I encountered through the book, on page and behind the scenes. Your inspiring viewpoints, work, and research deserve a greater amount of public support. For those wanting to get involved, here's a short list of groups trying to make the best of the Anthropocene:

Xerces Society for Invertebrate Conservation
E.O. Wilson Biodiversity Foundation
Monarch Join Venture
Bumblebee Conservation Trust
Honey Bee Health Coalition
The Entomological Foundation
Buglife (UK)
Detroit Zoo
The Nature Conservancy

Speaking of behind-the-scene folks, a HUGE THANKS to all the professionals commenting on early drafts of each chapter, suggesting ideas and catching boo-boos along the way: Robert Nathan Allen, aka RNA; James N. Hogue; Mohammad-Amir Aghaee; Vincent H. Resh; Adrian Smith; Michelle Sanford; Eric Benbow; Evan Paul Cherniack; Mario Padilla; and Elina L. Niño. As a journalist with no scientific background—besides coming from a family tree of engineers—diving into such a complex subject as entomology can be daunting. Your help was greatly appreciated. Infinite thank-yous, as well, to Gwen Pearson; she not only helped in finding commenters for the book, but happens to write incredibly fun and accessible articles about insects. Ditto the large number of great writers/entomologists doing the same, like May Berenbaum. If you liked this book, you will definitely love their work.

Besides the fascinating characters I met, there were people who took me in during my travels, hosting me and sharing great conversations—both strangers turned friends and friends turned family.

So, thank you: Sandro Batista for welcoming me into South America for the first time (sorry, again, about the ink blot–stained blanket); Adam Conrad and Dan Zamzow for the Minneapolis hospitality and finally exposing me to *Rick and Morty*; my Brooklyn/Burning Man family members Doug Bierend, Jenn "Flash" Salvemini, and Aaron Stone for giving me a bed and a decade of encouragement; Eiji Ohya for coming to the rescue during my *Lost in Translation* experience in Tokyo; Mushimoiselle Giriko for arranging that excellent Bug Crawl and her translator friend and talented illustrator Eri Sasayama; Ayomi Anabuki-Browning for organizing the field trip to the Tomioka Silk Mill; Andrea Cavallera and his wife, Laura Caracristi, for being such great hosts and drinking buddies in Brazil; Robyn Schoen for coming in clutch with a couch in Austin; my Airbnb host Diana from Brighton for the lead on the Hive at Kew Gardens; Rosie from Bristol for an amazing, albeit brief two-hour conversation, despite me sweating like "ice cream"; the kind and accommodating monks at Buckfast Abbey; Lina Tsingerlioti, Xenia Regina and Lefteris Karoutsos for helping me arrange my stay in Ikaria; and Aleksander

Sønder for the room in Copenhagen and the eye-opening experience at his lakeside cabin with the rest of the Danish crew. Skål!

For both sides of my family, the MacNeals and the Sururs: Thank you, habayyeb! (That's Arabic for "my loves.")

Thanks also to my friends Robert & Lauren Schauer, Jim & Tara Grude, Andrew & Christina Whyte, the Nyby Brothers, Nick Gutierrez, and Tasha Finken for participating in said Bug Feast and through the years making me laugh till I cried. Joining in with the laughs has been my *Big Lebowski*–obsessed friend Anne-Claire Siegert, as well as John & Ellie Nonemacher and Tom DeFreytas (really appreciate you aiding me in the cockroach surgeries). The coffee crew at Pablo's on 6th Ave. were consistently uplifting, especially Lauren. Same goes for my main bagel-slinger, Peter Paul Russo at Rosenberg's, for talking comic books and brightening my day with rugelach.

And, most of all, to my grandparents, who, as the Greeks say, planted "trees whose shade they know they will never sit in."

Bibliography

Chapter One: A Cabinet of Curiosity

Throughout this chapter, and for that matter the entire book, I have researched from two crucial sources for which I couldn't be more thankful: History of Entomology *and* Encyclopedia of Insects. *They both served as insect bibles, and a number of quotes come from these books. Additionally, J. F. M. Clark's* Bugs and the Victorians *was an excellent source.*

Aiso, Shigetoshi, et al. "Carcinogenicity and Chronic Toxicity in Mice and Rats Exposed by Inhalation to *Para*-dichlorobenzene for Two Years." *Journal of Veterinary Medical Science* 67:10 (2005): 1019–29.

Axson, Scooby. "New Campus Trend: Cricket Spitting?" Columbia News Service, April 4, 2012. http://www.startribune.com/new-campus-trend-cricket-spitting/146103835/

Bamford, Mary Ellen. *The Second Year of the Look-About Club.* Boston: Lothrop, Lee & Shepard Co., 1889.

Boisduval, Jean-Baptiste. "Notice: Sur M. Le Conte DeJean." *Annals of the Entomological Society of France* 2 (1845): 502–3.

Clark, J. F. M. *Bugs and the Victorians.* New Haven, CT: Yale University Press, 2009.

Dacke, Marie, et al. "Dung Beetles Use the Milky Way for Orientation." *Current Biology* 23:4 (2013): 298–300.

Damkaer, David M. *The Copepodologist's Cabinet: A Biographical and Bibliographical History.* Philadelphia: American Philosophical Society, 2002.

Epstein, Marc E., and Pamela M. Henson. "Digging for Dyar: The Man behind the Myth." *American Entomologist* 38:3 (1992): 148–71.

Gillott, Cedric. *Entomology.* New York: Plenum Press, 1980.

Häggqvist, Sibylle, Sven Olof Ulefors, and Fredrik Ronquist. "A New Species Group in *Megaselia*, the *Lucifrons* Group, with Descriptions of New Species." *ZooKeys* 512 (2015): 89–108.

Jones, Emma. *Tuscany and Umbria*. Edison, NJ: Hunter Publishing Inc., 2008.

Kaiser, Aaron. "Springfest Offers Family-Friendly Activities, Including Cricket Spitting." *The Exponent*, April 11, 2014. http://www.purdueexponent.org/features /article_1dccf9bc-3eb1-5e31-bdf9-bf6dbf75138a.html

Louis, Figuier, and Peter Martin Duncan. *The Insect World*. New York: D. Appleton, 1872.

Lubbock, John. "On the Objects of a Collection of Insects." *The Entomologist's Annual* (1856): 115–21.

Mawdsley, Jonathan R. "The Entomological Collection of Thomas Say." *Psyche* 100:3–4 (1993): 163–71.

Plautz, Jason. "Schwarzenegger Beetles (and Other Celebrity Species)." *Mental Floss*, July 29, 2008. http://mentalfloss.com/article/19203/schwarzenegger-beetles-and -other-celebrity-species

Resh, Vincent H., and Ring T. Cardé, eds. *Encyclopedia of Insects*. Boston: Academic Press, 2003.

Rothschild, Miriam, et al. "Execution of the Jump and Activity." *Philosophical Transactions of the Royal Society of London* 271:914 (1975): 499–515.

Schiebinger, Londa. *The Mind Has No Sex?: Women in the Origins of Modern Science*. Cambridge, MA: Harvard University Press, 1989.

Seto, Chris. "Guelph Barcoding Conference Highlights Need for 'Library of Life.'" *Guelph Mercury*, August 9, 2015. http://www.guelphmercury.com/news-story /5804236-guelph-barcoding-conference-highlights-need-for-library-of-life-/

Seven Wonders of the World: Miriam Rothschild. Narr. Sue Lawley. Dir. Christopher Sykes. BBC. 1995.

Shepardson, Daniel P. "Bugs, Butterflies, and Spiders: Children's Understandings about Insects." *International Journal of Science Education* 24:6 (2002): 627–43.

Smith, Ray F., Thomas E. Mittler, and Carroll N. Smith, eds. *History of Entomology*. Palo Alto, CA: Annual Reviews Inc., 1973.

Spilman, T. J. "Vignettes of 100 Years of the Entomological Society of Washington." *Proceedings of the Entomological Society of Washington* 86:1 (1984): 1–10.

Chapter Two: Buried Cities

This chapter would not exist were it not for the revolutionary work of E. O. Wilson and Bert Hölldobler. Pull quotes from The Ants *and* Journey to the Ants *were used for the "Gospel" inserts in the chapter, as was Mr. Wilson's fantastic novel* Anthill. *Quotes from Deborah M. Gordon were from both our interview and her impressive research papers and book.*

Anderson, David J., and Ralph Adolphs. "A Framework for Studying Emotions across Phylogeny." *Cell* 157:1 (2014): 187–200.

Bova, Jake. "Do Insects Feel Pain?" *Relax I'm an Entomologist*. Tumblr, May 25, 2013. http://relaximanentomologist.tumblr.com/post/51301520453/do-insects-feel-pain

Colorni, Alberto, Marco Dorigo, and Vittorio Maniezzo. "Distributed Ant Optimization by Ant Colonies." *Proceedings of the First European Conference on Artificial Life*. Elsevier Publishing (1991): 134–42.

de Bruyn, L. A. Lobry, and A. J. Conacher. "The Role of Termites and Ants in Soil Modification: A Review." *Australian Journal of Soil Research* 28 (1990): 55–93.

Dorn, Ronald. "Ants as Powerful Biotic Agent of Olivine and Plagioclase Dissolution." *Geology* 42:9 (2014): 771–74.

Drager, Kim. "Plasticity of Soil-Dwelling Ant Nest Architecture and Effects on Soil Properties in Environments of Contrasting Soil Texture." Entomological Society of America Conference, November 17, 2015, Minneapolis Convention Center, Minneapolis, Minn., Joint Symposium.

E. O. Wilson: Of Ants and Men. Writ. Graham Townsley. Dir. Shelly Schulze. Shining Red Productions, Inc., 2015. http://www.pbs.org/program/eo-wilson/

Gordon, Deborah M. *Ant Encounters: Interaction Networks and Colony Behavior.* Princeton, NJ: Princeton University Press, 2010.

——. "From Division of Labor to the Collective Behavior of Social Insects." *Behavioral Ecology and Sociobiology* (2015): 1–8.

——. "What Ants Teach Us About the Brain, Cancer and the Internet." TED 2014. https://www.ted.com/talks/deborah_gordon_what_ants_teach_us_about_the _brain_cancer_and_the_internet

Gorman, James. "To Study Aggression, A Fight Club for Flies." *New York Times*, February 3, 2014. https://www.nytimes.com/2014/02/04/science/to-study-aggression-a -fight-club-for-flies.html

Hölldobler, Bert, and Edward O. Wilson. *Journey to the Ants: A Story of Scientific Exploration.* Cambridge, MA: The Belknap Press of Harvard University Press, 1995.

——. *The Ants.* Cambridge, MA: The Belknap Press of Harvard University Press, 1990.

Hoy, Ron, and Jayne Yack. "Hearing." *Encyclopedia of Insects.* Vincent H. Resh and Ring T. Cardé, eds. 2nd edition. Boston: Academic Press, 2009. 440–46.

Ito, Kei, et al. "A Systematic Nomenclature for the Insect Brain." *Neuron* 81:4 (2014): 755–65.

Land, Michael F. "Eyes and Vision." *Encyclopedia of Insects.* Vincent H. Resh and Ring T. Cardé, eds. 2nd edition. Boston: Academic Press, 2009. 345–55.

Mery, Frederic, and Tadeusz J. Kawecki. "Experimental Evolution of Learning Ability in Fruit Flies." *Proceedings of the National Academy of Science* 99:22 (2002): 14274–79.

Mitchell, B. K. "Chemoreception." *Encyclopedia of Insects.* Vincent H. Resh and Ring T. Cardé, eds. 2nd edition. Boston: Academic Press, 2009. 148–52.

Moser, John C. "Contents and Structure of *Atta texana* Nest in Summer." *Annals of the Entomological Society of American* 56:3 (1963): 286–91.

Papaj, Daniel R., and Emilie C. Snell-Rood. "Memory Flies Sooner from Flies that Learn Faster." *Proceedings of the National Academy of Science* 104:34 (2007): 13539–40.

Planet Ant: Life Inside the Colony. Narr. George McGavin and Adam G. Hart. Dir. Graham Russell. BBC. 2012.

Prabhakar, Balaji, Katherine N. Dektar, and Deborah M. Gordon. "Anternet: The Regulation of Harvester Ant Foraging and Internet Congestion Control." *Communication, Control, and Computing, 2012 50th Annual Allerton Conference* (2012): 1355–59.

Schultz, Kevin. "Ants May Boost CO_2 Absorption Enough to Slow Global Warming." *Scientific American*, August 12, 2014. https://www.scientificamerican.com/article /ants-may-boost-co2-absorption-enough-to-slow-global-warming/

Sierzputowski, Kate. "Macro Photograph's of Nature's Tiniest Architects by Nick Bay." *This Is Colossal*, February 19, 2016. http://www.thisiscolossal.com/2016/02/macro -photographs-of-natures-tiniest-architects-by-nicky-bay/

Strausfeld, Nicholas J. "Brain and Optic Lobes." *Encyclopedia of Insects.* Vincent H. Resh and Ring T. Cardé, eds. 2nd edition. Boston: Academic Press, 2009. 121–30.

Stützle, Thomas, Manuel López-Ibáñez, and Marco Dorigo. "A Concise Overview of Applications of Ant Colony Optimization." *Encyclopedia of Operations Research and Management Science.* Hoboken, NJ: John Wiley & Sons, 2011.

Tompkins, Joshua. "Empire of the Ant." *Los Angeles Magazine*, February 2001, 66.

Tschinkel, Walter R. "Subterranean Ant Nests: Trace Fossils Past and Future?" *Palaeogeography, Palaeoclimatology, Palaeoecology* 192 (2003): 321–33.

"Visualizing the Future." *State of Tomorrow.* The University of Texas Foundation. PBS. 2012.

Wangberg, James K. *Do Bees Sneeze?: And Other Questions Kids Ask About Insects.* Golden, CO: Fulcrum Publishing, 1997.

Wilson, Edward O. *Anthill.* New York: W.W. Norton & Company, 2010.

——. *Sociobiology: The New Synthesis.* Cambridge, MA: The Belknap Press of Harvard University Press, 2000.

Wilson, Mark. "An Ant Ballet, Choreographed by Pheromones and Robots." *Fast Company's Co. Design*, May 23, 2012. https://www.fastcodesign.com/1669858/an-ant-ballet-choreographed-by-pheromones-and-robots

Yong, Ed. "Ants Write Architectural Plans into the Walls of Their Buildings." *National Geographic*, January 18, 2016. http://phenomena.nationalgeographic.com/2016/01/18/ants-write-architectural-plans-into-the-walls-of-their-buildings/

——. "Tracking Whole Colonies Shows Ants Make Career Moves." *Nature*, April 18, 2013. http://www.nature.com/news/tracking-whole-colonies-shows-ants-make-career-moves-1.12833

Chapter Three: "Even Educated Fleas Do It"

Insect reproduction and, more so, conservation are little-discussed imperatives of the world. At least that's what I found, and why these works and theories were extremely helpful during the research for this chapter. Robert Dunn's paper and the works of Michael Samways (Insect Diversity Conservation and Insect Conservation) are largely featured, as well as Jeffrey Lockwood's American Entomologist paper titled "Voices from the Past" and his book The Infested Mind. As for the intriguing subject of bug sex, I recommend James Wangberg's Six-Legged Sex and Marlene Zuk's Sex on Six Legs, which both shed a lot of light on bizarre sexual acts. And speaking of Marlene Zuk, our interview was very illuminating, as was my interview with Timothy Mousseau.

Adamo, Shelley A., et al. "Climate Change and Temperate Zone Insects: The Tyranny of Thermodynamics Meets the World of Limited Resources." *Environmental Entomology* 41:6 (2012): 1644–52.

Angilletta, Michael J., Jr., Raymond B. Huey, and Melanie R. Frazier. "Thermodynamic Effects on Organismal Performance: Is Hotter Better?" *Physiological and Biochemical Zoology* 83:2 (2010): 197–206.

Associated Press. "Man Arrested for Lighting Tarantula-Fed Fire." *Lubbock Avalanche-Journal*, July 20, 2003. http://lubbockonline.com/stories/072003/reg_0720030077.shtml#.WIaQ3bYrKi4

Berenbaum, May R. *Bugs in the System: Insects and Their Impact on Human Affairs.* Boston: Addison-Wesley, 1995.

——. "Rad Roaches." *American Entomologist* 47:3 (2001): 132–33.

Binks, S., D. Chan, and N. Medford. "Abolition of Lifelong Specific Phobia: A Novel Therapeutic Consequence of Left Mesial Temporal Lobectomy." *Neurocase* 21:1 (2015): 79–84.

Bonebrake, Timothy C., and Curtis A. Deutsch. "Climate Heterogeneity Modulates Impact of Warming on Tropical Insects." *Ecology* 93:2 (2012): 449–55.

Boyd, Brian, and Robert Michael Pyle, eds. *Nabokov's Butterflies*. Boston: Beacon Press, 2000.

Caballero-Mendieta, N., and Carlos Cordero. "Enigmatic Liaisons in Lepidoptera: A Review of Same-Sex Courtship and Copulation in Butterflies and Moths." *Journals of Insect Science* (2012) 12:138. Available online: http://www.insectscience.org/12.138.

Choe, Jae C., and Bernard J. Crespi, eds. *The Evolution of Mating Systems in Insects and Arachnids*. Cambridge: Cambridge University Press, 1997.

Dewaraja, Ratnin. "Formicophilia, an Unusual Paraphilia, Treated with Counseling and Behavior Therapy." *American Journal of Psychotherapy* 41:4 (1987): 593–97.

Dunn, Robert R. "Modern Insect Extinctions, the Neglected Majority." *Conservation Biology* 19 (2005): 1030–36.

Fountain, Henry. "At Chernobyl, Hints of Nature's Adaptation." *New York Times*, May 5, 2014. https://www.nytimes.com/2014/05/06/science/nature-adapts-to-chernobyl.html

Giant weta/wetapunga. New Zealand's Department of Conservation. http://www.doc.govt.nz/nature/native-animals/invertebrates/weta/giant-weta-wetapunga/

Hajna, Larry. "Biologist Studies Elusive Worm." *Courier-Post*, March 28, 2006. https://beta.groups.yahoo.com/neo/groups/ParanormalGhostSociety/conversations/messages/36043

Hogue, Charles Leonard. *Latin American Insects and Entomology*. Berkeley: University of California Press, 1993.

Kaplan, Sarah. "The White House Plan to Save the Monarch Butterfly: Build a Butterfly Highway." *Washington Post*, May 21, 2015. https://www.washingtonpost.com/news/morning-mix/wp/2015/05/21/the-white-house-plan-to-save-the-monarch-butterfly-build-a-butterfly-highway/?utm_term=.6d1c902fb9d7

Kritsky, Gene, and Ron Cherry. *Insect Mythology*. San Jose, CA: Writers Club Press, 2000.

Krulwich, Robert. "Six-Legged Giant Finds Secret Hideaway, Hides for 80 Years." *NPR*, February 29, 2012. http://www.npr.org/sections/krulwich/2012/02/24/147367644/six-legged-giant-finds-secret-hideaway-hides-for-80-years

Li, Shu, et al. "Forever Love: The Hitherto Earliest Record of Copulating Insects from the Middle Jurassic of China." *PLoS ONE* (2013): e78188.

Lloyd, J. E. "Mating Behavior and Natural Selection." *Florida Entomologist* 62:1 (1979): 17–34.

Lockwood, Jeffery A. "Voices from the Past: What We Can Learn from the Rocky Mountain Locust." *American Entomologist* 47:4 (2001): 208–15.

——. *The Infested Mind: Why Humans Fear, Loathe, and Love Insects*. New York: Oxford University Press, 2013.

Møller, Anders, and Timothy A. Mousseau. "Reduced Abundance of Insects and Spiders Linked to Radiation at Chernobyl 20 Years after the Accident." *Biology Letters* 5:3 (2009): 356–59.

Molur, Sanjay, Manju Silliwal, and B. A. Daniel. "At Last! Indian Tarantulas on ICUN Red List." *Zoo's Print* 506 (2008): 1–3.

O'Brien, R. D., and L. S. Wolfe. *Radiation, Radioactivity, and Insects*. New York: Academic Press, 1964.

Paynter, Ben. "The Bug Wrangler." *Wired*, May 2012: 113.

Penny, D., and J. E. Jepson. *Fossil Insects: An Introduction to Palaeoentomology*. Manchester, UK: Siri Scientific Press, 2014.

Priddel, David, et al. "Rediscovery of the 'Extinct' Lord Howe Island Stick Insect and Recommendations for Its Conservation." *Biodiversity and Conservation* 12 (2003): 1391–1403.

Pyle, Robert Michael. "Between Climb and Cloud." *Nabokov's Butterflies: Unpublished and Uncollected Writings.* Brian Boyd and Robert Michael Pyle, eds. Boston: Beacon Press, 2000.

"Quick Evolution Leads to Quiet Crickets." *Understanding Evolution.* University of California, Berkeley. http://evolution.berkeley.edu/evolibrary/news/061201 _quietcrickets

Ricciuti, Ed. "Wisconsin Butterfly Conservation Program Could Be Model for Future Efforts." *Entomology Today,* Entomological Society of America, June 26, 2015. https://entomologytoday.org/2015/06/26/wisconsin-butterfly-conservation-program -could-be-a-model-for-future-efforts/

Rothenberg, David. *Bug Music: How Insects Gave Us Rhythm and Noise.* New York: St. Martin's Press, 2013.

Sadowski, Jennifer A., Allen J. Moore, and Edmund D. Brodie III. "The Evolution of Empty Nuptial Gifts in a Dance Fly, *Empis snoodyi:* Bigger Isn't Always Better." *Behavioral Ecological Sociobiology* 45 (1999): 161–66.

Samways, Michael J. *Insect Diversity Conservation.* Cambridge: Cambridge University Press, 2005.

——. "Insect Extinctions and Insect Survival." *Conservation Biology* 20:1 (2006): 245–46.

Samways, Michael J., Melodie A. McGeoch, and Tim R. New. *Insect Conservation: A Handbook of Approaches and Methods.* New York: Oxford University Press, 2010.

Schultz, Stanley A., and Marguerite J. Schultz. *The Tarantula Keeper's Guide.* Hauppauge, NY: Barron's, 1998.

Schwander, Tanja, Lee Henry, and Bernard J. Crespi. "Molecular Evidence for Ancient Asexuality in *Timema* Stick Insects." *Current Biology* 21:13 (2011): 1129–34.

Shain, Daniel H. "The Ice Worm's Secret." *Alaska Park Science Journal* 3:1 (2004): 31.

United Press International. "National Insect: Bee or Butterfly?" *Lodi News-Sentinel,* December 13, 1989: 10.

Taira, Wataru, et al. "Fukushima's Biological Impacts: The Case of the Pale Grass Blue Butterfly." *Journal of Heredity* 105:5 (2014): 710–22.

Tinghitella, R.M., et al. "Island Hopping Introduces Polynesian Field Crickets to Novel Environments, Genetic Bottlenecks and Rapid Evolution." *Journal of Evolutionary Biology* 24 (2011): 1199–1211.

Wangberg, James K. *Six-Legged Sex: The Erotic Lives of Bugs.* Golden, CO: Fulcrum Publishing, 2001.

Whitcomb, W. H., and R. Eason. "The Mating Behavior of *Peucetia viridans*." *Florida Entomologist* 48:3 (1964): 163–67.

Wilson, Edward O. "The Little Things that Run the World (the Importance and Conservation of Invertebrates)." *Conservation Biology* 1 (1987): 344–46.

Yoshizawa, Kazunori, et al. "Female Penis, Male Vagina, and Their Correlated Evolution in a Cave Insect." *Current Biology* 24:9 (2014): 1006–10.

Zuk, Marlene. *Sex on Six Legs: Lessons on Life Love & Language from the Insect World.* New York: Houghton Mifflin Harcourt, 2011.

——. "What We Learn from Insects' Sex Lives." TED Women 2015. https://www.ted.com /talks/marlene_zuk_what_we_learn_from_insects_kinky_sex_lives

Zuk, Marlene, John T. Rotenberry, and Robin M. Tinghitella. "Silent Night: Adaptive Disappearance of a Sexual Signal in a Parasitized Population of Field Crickets." *Biology Letters* 2:4 (2006): 521–24.

Chapter Four: The On-Flying Things

Hands down the two books taking the spotlight in researching for this chapter were Molly Caldwell Crosby's The American Plague *and Jim Murphy's* An American Plague. *Similarly titled but distinctly different in their narrative focus on yellow fever epidemics. Ditto the books* Justinian's Flea *and* Rats, Lice and History, *in case you want to know more about arboviruses. My interview with Mike Turell was especially informative. As for the ecological impact of beetles, Andrew Nikiforuk's* Empire of the Beetle *served as another main source for the chapter.*

Alexander, Renée. "Engineering Mosquitoes to Spread Health." *The Atlantic*, September 14, 2014. http://www.theatlantic.com/health/archive/2014/09/engineering -mosquitoes-to-stop-disease/379247/

Alvarez, Lizette. "Citrus Disease with No Cure Is Ravaging Florida Groves." *New York Times*, May 9, 2013. http://www.nytimes.com/2013/05/10/us/disease-threatens -floridas-citrus-industry.html

Barnabas Health. Clara Maass Medical Center, History, n.p. http://www.barnabashealth .org/Clara-Maass-Medical-Center/About-Us/History.aspx

Bar-Zeev, Micha, and Rachel Galun. "A Magnetic Method of Separating Mosquito Pupae from Larvae." *Mosquito News* 21:3 (1961): 225–28.

Berntson, Ben. "Boll Weevil Monument." *Encyclopedia of Alabama*, June 7, 2013. http:// www.encyclopediaofalabama.org/article/h-2384

Brunton, Sir Lauder. "Fleas as a National Danger." *Journal of Tropical Medicine and Hygiene* 10 (1907): 388–91.

Carey, Matt. *The History of Vaccines.* The College of Physicians of Philadelphia, July 7, 2010. http://www.historyofvaccines.org/

Crosby, Molly Caldwell. *The American Plague: The Untold Story of Yellow Fever, the Epidemic that Shaped Our History.* New York: Berkley Books, 2006.

Davis, Simon. "Solving the Mystery of an Ancient Epidemic." *The Atlantic*, September 15, 2015. http://www.theatlantic.com/health/archive/2015/09/disease-plague-of -athens-ebola/403561/

de Valdez, Megan R. Wise, et al. "Genetic Elimination of Dengue Vector Mosquitoes." *Proceedings of the National Academy of Sciences* 108:12 (2011): 4772–75.

Dewar, Heather. "Did Mosquito Bite Defeat Alexander?" *Baltimore Sun*, December 13, 2003. http://articles.baltimoresun.com/2003-12-13/news/0312130076_1_nile-virus -west-nile-alexander

Elsevier. "Typhoid Fever Led to the Fall of Athens." *ScienceDaily*, January 23, 2006. https://www.sciencedaily.com/releases/2006/01/060123163827.htm

Enserink, Martin. "GM Mosquito Trial Alarms Opponents, Strains Ties in Gates-Funded Project." *Science* 330 (2010): 1030–31.

Finlay, Charles. "The Mosquito Hypothetically Considered as an Agent in the Transmission of Yellow Fever Poison." *New Orleans Medical and Surgical Journal* 9 (1882): 601–16.

Frith, John. "The History of Plague Pt. 2. The Discoveries of the Plague Bacillus and Its Vector." *Journal of Military and Veteran's Health* 20:3 (2012). Web.

Funk, Jason, and Stephen Saunders. *Rocky Mountain Forests at Risk: Confronting Climate-Driven Impacts from Insects, Wildfires, Heat, and Drought.* Union of Concerned Scientists and the Rocky Mountain Climate Organization, 2014. http://www

.ucsusa.org/sites/default/files/attach/2014/09/Rocky-Mountain-Forests-at-Risk
-Full-Report.pdf

Gladwell, Malcolm. "The Mosquito Killer." *New Yorker,* July 2, 2001. http://www.newyorker
.com/magazine/2001/07/02/the-mosquito-killer

Glick, P.A. *The Distribution of Insects, Spiders and Mites in the Air.* Washington, DC:
US Department of Agriculture, 1939. https://naldc.nal.usda.gov/naldc/download
.xhtml?id=CAT86200667&content=PDF

Guidi, Rodrigo. "*Aedes aegypti* do Bem reduz em 82 percent as larvas selvagens do mosquito
no Cecap/Eldorado." *Prefeitura do Município de Piracicaba.* Secretaria Municipal da
Saúde, January 19, 2016. http://www.piracicaba.sp.gov.br/aedes+aegypti+do+bem+red
uz+em+82+as+larvas+selvagens+do+mosquito+no+cecap+eldorado.aspx

Hofstetter, Richard W., et al. "Using Acoustic Technology to reduce Bark Beetle Reproduction." *Pest Management Science* 70:1 (2014): 24–27.

Hogue, James N. "Insects Effect on Human History." *Encyclopedia of Insects.* Vincent
H. Resh and Ring T. Cardé, eds. 2nd edition. Boston: Academic Press, 2009.
471–73.

Keim, Brandon. "Marvelous Destroyers: The Fungus-Farming Beetle." *Wired,* July 27,
2011. https://www.wired.com/2011/07/fungus-farming-beetles/

Konkel, Lindsey. "Invasion of the Pine Beetles." *OnEarth,* October 5, 2009. http://archive
.onearth.org/article/invasion-of-the-pine-beetles

Littman, Robert J. "The Plague of Athens: Epidemiology and Paleopathology." *Mount
Sinai Journal of Medicine* 76 (2009): 456–67.

Logan, Jesse A., and James A. Powell. "Ghost Forests, Global Warming, and the Mountain Pine Beetle." *American Entomologist* 47:3 (2001): 160–73.

"Malaria Facts." *About Malaria.* Centers for Disease Control and Prevention. https://
www.cdc.gov/malaria/about/facts.html

Millar, Constance I., Robert D. Westfall, and Diane L. Delany. "Response of High-
Elevation Limber Pine to Multiyear Droughts and 20th-Century Warming,
Sierra Nevada, California, USA." *Canadian Journal of Forest Research* 37:12 (2007):
2508–20.

Murphy, Jim. *An American Plague: The True and Terrifying Story of the Yellow Fever
Epidemic of 1793.* New York: Clarion Books, 2003.

Netburn, Deborah. "Scientists Aim to Fight Malaria with Genetically Engineered Mosquitoes." *Los Angeles Times,* November 25, 2015. http://www.latimes.com/science
/sciencenow/la-sci-sn-genetically-engineered-mosquitoes-malaria-20151121-story
.html

Nikiforuk, Andrew. *Empire of the Beetle: How Human Folly and a Tiny Bug Are Killing
North America's Great Forests.* Vancouver: Greystone Books, 2011.

Novy, James E. "Screwworm Control and Eradication in the Southern United States of
America." *World Animal Review* (1991): 18–27.

Oatman, Maddie. "Bark Beetles Are Decimating Our Forests. That Might Be a Good
Thing." *Mother Jones,* March 19, 2015. http://www.motherjones.com/environment
/2015/03/bark-pine-beetles-climate-change-diana-six

Reed, Walter, James Carroll, and Aristides Agramonte. "The Etiology of Yellow Fever:
An Additional Note." *Journal of the American Medical Association* 36:7 (1901):
431–40.

Rosen, William. *Justinian's Flea: Plague, Empire, and the Birth of Europe.* New York:
Viking, 2007.

Rosner, Hillary. "The Bug that's Eating the Woods." *National Geographic*, April 2015. http://ngm.nationalgeographic.com/2015/04/pine-beetles/rosner-text

Shapiro, Beth, Andrew Rambaut, and M. Thomas P. Gilbert. "No Proof that Typhoid Caused the Plague of Athens (a Reply to Papagrigorakis et al.)." *International Journal of Infectious Diseases* 10:4 (2006): 334–35.

Six, Diana L., Eric Biber, and Elisabeth Long. "Management for Mountain Pine Beetle Outbreak Suppression: Does Relevant Science Support Current Policy?" *Forests* 5:1 (2014): 103–33.

Soupios, M. A. "Impact of the Plague in Ancient Greece." *Infectious Disease Clinics of North America* 18 (2004): 45–51.

Specter, Michael. "The Mosquito Solution." *New Yorker*, July 9, 2012. http://www.newyorker.com/magazine/2012/07/09/the-mosquito-solution

Thucydides. *History of the Peloponnesian War*. Trans. Richard Crawley. New York: Dutton, 1950.

Wade, Nicholas. "Engineering Mosquitoes' Genes to Resist Malaria." *New York Times*, November 23, 2015. https://www.nytimes.com/2015/11/24/science/gene-drive-mosquitoes-malaria.html

Waltz, Emily. "Oxitec Trials GM Sterile Moth to Combat Agricultural Infestations." *Nature Biotechnology* 33 (2015): 792–793.

Westerling, Anthony L., et al. "Continued Warming Could Transform Greater Yellowstone Fire Regimes by Mid-21st Century." *Proceedings of the National Academy of Sciences* 108:32 (2011): 13165–70.

Wilford, John Noble. "DNA Shows Malaria Helped Topple Rome." *New York Times*, February 20, 2001. http://www.nytimes.com/2001/02/20/science/dna-shows-malaria-helped-topple-rome.html

Winslow, Charles-Edward A. *The Conquest of Epidemic Disease: A Chapter in the History of Ideas*. Princeton, NJ: Princeton University Press, 1944.

Zinsser, Hans. *Rats, Lice and History*. New York: Blue Ribbon Books, 1934.

Chapter Five: Vámonos Pest!

The pest control industry is a business world one does not often think of, so there are a couple sources that were extremely helpful in research and that heavily contributed to this chapter. Kicking it off was Mark Winston's Nature Wars. This is followed closely by Dawn Day Biehler's Pests in the City and Brooke Borel's Infested. Michael F. Potter's American Entomologist paper, "The History of Bed Bug Management," was also a key resource, and helping on the chemical control end as well was Will Allen's The War on Bugs.

Allen, Will. *The War on Bugs*. White River Junction, VT: Chelsea Green Publishing, 2008.

American Public Health Association. "Typhoid Fever Death Rates." *American Journal of Public Health* 13:8 (1923): 660.

Barre, H. W., and A. F. Conradi. *Treatment of Plant Diseases and Injurious Insects in South Carolina*. Columbia, SC: The R.L. Bryan Company, 1909.

Bed Bug TV. "Killing Bed Bugs: Steam vs. Cryonite." Online video clip. *YouTube*. February 23, 2012. https://www.youtube.com/watch?v=S9Qgf2EC358

Benoit, Joshua B., et al. "Unique Features of a Global Human Ectoparasite Identified through Sequencing of the Bed Bug Genome." *Nature Communications* 7 (2016). http://www.nature.com/articles/ncomms10165

Biehler, Dawn Day. *Pests in the City: Flies, Bedbugs, Cockroaches, and Rats*. Seattle: University of Washington Press, 2013.

Borel, Brooke. *Infested: How the Bed Bug Infiltrated Our Bedrooms and Took Over the World*. Chicago: University of Chicago Press, 2015.

Buckley, Cara. "Doubts Rise on Bedbug-Sniffing Dogs." *New York Times*, November 11, 2010.

" 'Bug Battle' at White House Is Won." *Washington Times*, September 8, 1924.

Ceccatti, John S. "Insecticide Resistance, Economic Entomology, and the Evolutionary Synthesis, 1914–1951." *Transactions of the American Philosophical Society* 99:1 (2009): 199–217.

Cowan, Robin, and Philip Gunby. "Sprayed to Death: Path Dependence, Lock-in and Pest Control Strategies." *Economic Journal* 106 (1996): 521–42.

"Don't Let the Bed Bugs Bite Act of 2009." H.R. 2248, 111th Congress. (2009). https://www.congress.gov/bill/111th-congress/house-bill/2248

Environmental Protection Agency. *DDT Ban Takes Effect*. December 31, 1972. https://www.epa.gov/history/epa-history-ddt-dichloro-diphenyl-trichloroethane

Genzlinger, Neil. "That Itchy Feeling? It May Just Be Love." *New York Times*, September 15, 2014. https://www.nytimes.com/2014/09/16/theater/bedbugs-the-musical-is-at-the-arclight-theater.html

Gorman, James. "Wily Cockroaches Find Another Survival Trick: Laying Off the Sweets." *New York Times*, May 23, 2013. http://www.nytimes.com/2013/05/24/science/a-bitter-sweet-shift-in-cockroach-defenses.html

Harbison, Brad. "Bed Bugs in NYC: One PMP's Perspective." *Pest Control Technology*, January 14, 2016. http://www.pctonline.com/article/bed-bugs-in-nyc—one-pmps-perspective/

Hoddle, M. S., and R. G. Van Driesche. "Biological Control of Insect Pests." *Encyclopedia of Insects*. Vincent H. Resh and Ring T. Cardé, eds. Boston: Academic Press, 2009. 91–101.

Horowitz, A. Rami, and Isaac Ishaaya, eds. *Advances in Insect Control and Resistance Management*. Cham, Switzerland: Springer International Publishing, 2016.

"How Does *Bt* Work?" *Bacillus thuringiensis*. University of California, San Diego. http://www.bt.ucsd.edu/how_bt_work.html

Hoyt, Erich, and Ted Schultz, eds. *Insect Lives: Stories of Mystery and Romance from a Hidden World*. Cambridge, MA: Harvard University Press, 1999.

Maia, Marta Ferreira, and Sarah J. Moore. "Plant-Based Insect Repellents: A Review of Their Efficacy, Development and Testing." *Malaria Journal* 10. Suppl. 1 (2011).

Maredia, K. M., D. Dakouo, and D. Mota-Sanchez, eds. *Integrated Pest Management in the Global Arena*. Cambridge, MA: CABI Publishing, 2003.

Mastalerz, Przemyslaw. *The True Story of DDT, PCB, and Dioxin*. Wroclaw, Poland: Wydawnictwo Chemiczne, 2005.

McWilliams, J. E. " 'The Horizon Opened up Very Greatly': Leland O. Howard and the Transition to Chemical Insecticides in the United States, 1894–1927." *Agricultural History* 82:4 (2008): 468–95.

New York Times Service. "A Short But Sweet History of the Flypaper Industry." *Milwaukee Journal*, October 22, 1976.

New York vs. Bed Bugs. Renee Corea, 2012. http://newyorkvsbedbugs.org/

Oliver, Simon, and Tony Moore. *The Exterminators. Bug Brothers*. New York: DC Comics, 2006.

"Pesticide Disaster in Mississippi." *Green Left Weekly*, March 26, 1997. https://www.greenleft.org.au/content/pesticide-disaster-mississippi

Potter, Michael F. "The History of Bed Bug Management: With Lessons from the Past." *American Entomologist* 57:1 (2011): 14–25.

The Rhythm Club Fire: A Documentary. Dir. Bryan Burch. 2010. https://www.youtube.com/watch?v=gXKxt3Ki6Lo

Rice, F. L., et al. "Crystalline Silica Exposure and Lung Cancer Mortality in Diatomaceous Earth Industry Workers: A Quantitative Risk Assessment." *Occupational and Environment Medicine* 58 (2001): 38–45.

Rosenfeld, Jeffrey A., et al. "Genome Assembly and Geospatial Phylogenomics of the Bed Bug *Cimex lectularius*." *Nature Communications* 7 (2016). http://www.nature.com/articles/ncomms10164

Schechner, Sam. "The Roach that Failed." *New York Times Magazine*, July 25, 2004. http://www.nytimes.com/2004/07/25/magazine/the-way-we-live-now-7-25-04-phenomenon-the-roach-that-failed.html

Smith, Allan E., and Diane M. Secoy. "Forerunners of Pesticides in Classical Greece and Rome." *Journal of Agricultural Food Chemistry* 23:6 (1975): 1050–55.

Sorenson, W. Conner, et al. "Charles V. Riley, France, and *Phylloxera*." *American Entomologist* 54:3 (2008): 134–49.

"SPC Report: U.S. Structural Pest Control Market Approaches $7.5 Billion." *Pest Control Technology Magazine*, April 10, 2015. http://www.pctonline.com/article/spc-2014-pest-control-market-report/

Tabashnik, Bruce E., et al. "Defining Terms for Proactive Management of Resistance to *Bt* Crops and Pesticides." *Journal of Economic Entomology* 107:2 (2014): 496–507.

Wang, Changlu, et al. "Bed Bugs: Prevalence in Low-Income Communities, Resident's Reactions, and Implementation of a Low-Cost Inspection Protocol." *Journal of Medical Entomology* (2016): 1–8.

Winston, Mark L. *Nature Wars: People vs. Pests*. Cambridge, MA: Harvard University Press, 1997.

Yamano, Yuko, Jun Kagawa, and Rumiko Ishizu. "Two Cases of Methyl Bromide Poisoning in Termite Exterminators." *Journal of Occupational Health* 43:5 (2001): 291–94.

Chapter Six: First Responders

For more riveting anecdotes involving bugs and dead bodies, I highly recommend Lee Goff's A Fly for the Prosecution *and these textbooks on the matter:* Entomology and the Law, Insect Evidence, Forensic Science, *and* Forensic Entomology. *They came in handy for finding material and shedding light on this macabre and vital scientific underground. On the other side of decomposition, John E. Losey and Mace Vaughan's paper really ignited the fire for this chapter; and as far as bug behavior goes, again,* Encyclopedia of Insects *came to the rescue.*

Anderson, Gail S. "Forensic Entomology." *Forensic Science: An Introduction to Scientific and Investigative Techniques*. Stuart H. James and Jon J. Nordby, eds. 2nd edition. Boca Raton, FL: Taylor & Francis, 2005.

Byrd, Jason H., and James L. Castner. *Forensic Entomology: The Utility of Arthropods in Legal Investigations*. 2nd edition. Boca Raton, FL: Taylor & Francis, 2010.

Dung Beetle Program. CSIRO, CSIROpedia, n.p. https://csiropedia.csiro.au/dung-beetle-program/

Dung Down Under: A Study in Biological Control. Dir. Roger Seccombe. CSIRO, 1972. Documentary.

Erzinçlioglu, Zakaria. *Maggots, Murder, and Men: Memories and Reflections of a Forensic Entomologist.* New York: Thomas Dunne Books, 2000.

Goff, M. Lee. *A Fly for the Prosecution: How Insect Evidence Helps Solve Crimes.* Cambridge, MA: Harvard University Press, 2000.

Greenberg, Bernard, and John Charles Kunich. *Entomology and the Law: Flies as Forensic Indicators.* Cambridge: Cambridge University Press, 2002.

Hanski, Ilkka. "Nutritional Ecology of Dung- and Carrion-Feeding Insects." Frank Slansky Jr. and J. G. Rodriguez, eds. New York: John Wiley & Sons, 1987.

Harris County Institute of Forensic Sciences. (2013). Autopsy Reports. Houston, TX. https://ifs.harriscountytx.gov/Pages/default.aspx

Kintz, Pascal, et al. "Fly Larvae: A New Toxicological Method of Investigation in Forensic Medicine." *Journal of Forensic Sciences* 35:1 (1990): 204–7.

Lockwood, Jeffrey A. "Insects as Weapons of War, Terror, and Torture." *Annual Review of Entomology* 57 (2012): 205–27.

Losey, John E., and Mace Vaughan. "The Economic Value of Ecological Services Provided by Insects." *BioScience* 56:4 (2006): 311–23.

Lynch, Heather J., and Paul R. Moorcroft. "A Spatiotemporal Ripley's K-Function to Analyze Interactions between Spruce Budworm and Fire in British Columbia, Canada." *Canadian Journal of Forest Research* 38 (2008): 3112–19.

MacIvor, J. Scott, and Andrew E. Moore. "Bees Collect Polyurethane and Polyethylene Plastics as Novel Nest Materials." *Ecosphere* 4:12 (2013): 155.

Martin, Michael. *Insect Evidence.* Mankato, MN: Capstone Press, 2007.

Nguyen, Trinh T. X., Jeffery K. Tomberlin, and Sherah Vanlaerhoven. "Ability of Black Soldier Fly Larvae to Recycle Food Waste." *Environmental Entomology* 44:2 (2015): 406–10.

Payne, Jerry A. "A Summer Carrion Study of the Baby Pig *Sus scrofa* Linnaeus." *Ecology* 46:5 (1965): 592–602.

Ridsdill-Smith, James, and Leigh W. Simmons. "Dung Beetles." *Encyclopedia of Insects.* Vincent H. Resh and Ring T. Cardé, eds. 2nd edition. Boston: Academic Press, 2009. 304–7.

Smolka, Jochen, et al. "Dung Beetles Use Their Dung Ball as a Mobile Thermal Refuge." *Current Biology* 22:20 (2012): R863–R864.

Syamsa, R. A., et al. "Forensic Entomology of High-Rise Buildings in Malaysia: Three Case Reports." *Tropical Biomedicine* 32:2 (2015): 291–99.

Waldbauer, Gilbert. *What Good Are Bugs?: Insects in the Web of Life.* Cambridge, MA: Harvard University Press, 2003.

Whitten, Max. "How One Man's Beetle-Mania Lorded It Over the Flies." *Sydney Morning Herald*, May 30, 2014. http://www.smh.com.au/comment/obituaries/how-one-mans-beetlemania-lorded-it-over-the-flies-20140529-zrrmp.html

Yang, Yu, et al. "Biodegradation and Mineralization of Polystyrene and Plastic-Eating Mealworms: Part 1. Chemical and Physical Characterization and Isotopic Tests." *Environmental Science & Technology* 49 (2015): 12080–86.

Chapter Seven: You Just Squashed the Cure for Cancer

Compared to other bug topics, research about the medical and robotic benefits of bugs is scarce. That's why for this chapter I must highlight just how useful several sources were.

Aaron T. Dossey's Natural Products Reports *paper called "Insects and Their Chemical Weaponry" and E. Paul Cherniack's* Alternative Medicine Review *papers "Bugs as Drugs," parts one and two, were main go-to sources. May R. Berenbaum's book* The Earwig's Tail *was helpful here and throughout the book, as were many of her columns and much of her research. And Jay Harman's book* The Shark's Paintbrush *was a nice find as there isn't much out there related to biomimicry.*

Berenbaum, May R. *The Earwig's Tail: A Modern Bestiary of Multi-Legged Legends.* Cambridge, MA: Harvard University Press, 2009.

Bozkurt, Alper, Robert F. Gilmour, and Amit Lal. "Balloon-Assisted Flight of Radio-Controlled Insect Biobots." *IEEE Transactions on Biomedical Engineering* 56:9 (2009): 2304–7.

Cherniack, E. Paul. "Bugs as Drugs, Part 1: Insects. The 'New' Alternative Medicine for the 21st Century?" *Alternative Medicine Review* 15:2 (2010): 124–35.

——. "Bugs as Drugs, Part 2: Worms, Leeches, Scorpions, Snails, Ticks, Centipedes, and Spiders." *Alternative Medicine Review* 16:1 (2011): 50–58.

Chiadini, Francesco, et al. "Insect Eyes Inspire Improved Solar Cells." *Optics and Photonics News* 22:4 (2011): 38–43.

Cohen, David. "Painless Needle Copies Mosquito's Stinger." *New Scientist*, April 4, 2002. https://www.newscientist.com/article/dn2121-painless-needle-copies-mosquitos -stinger/

Cornwell, P. B. *The Cockroach: A Laboratory Insect and an Industrial Pest.* London: Hutchinson, 1968.

Cowan, Frank. *Curious Facts in the History of Insects, Including Spiders and Scorpions.* Philadelphia: J.B. Lippincott & Co., 1865.

Cruse, Holk. "Robotic Experiments on Insect Walking." *Artificial Ethology.* Owen Holland and David McFarland, eds. New York: Oxford University Press, 2001.

Dardevet, Lucie, et al. "Chlorotoxin: A Helpful Natural Scorpion Peptide to Diagnose Glioma and Fight Tumor Invasion." *Toxins* 7:4 (2015): 1079–1101.

Dossey, Aaron T. "Insects and Their Chemical Weaponry: New Potential for Drug Discovery." *Natural Products Reports* 27:12 (2010): 1737–57.

Eisner, Thomas. *For Love of Insects.* Cambridge, MA: The Belknap Press of Harvard University Press, 2003.

Eldor, A., M. Orevi, and M. Rigbi. "The Role of Leech in Medical Therapeutics." *Blood Reviews* 10 (1990): 201–9.

Full, Robert. "The Secrets of Nature's Grossest Creatures, Channeled into Robots." TED, March 2014. https://www.ted.com/talks/robert_full_the_secrets_of_nature_s _grossest_creatures_channeled_into_robots

Graule, M. A., et al. "Perching and Takeoff of a Robotic Insect on Overhangs Using Switchable Electrostatic Adhesion." *Science* 352:6288 (2016): 978–82.

Harman, Jay. *The Shark's Paintbrush: Biomimicry and How Nature Is Inspiring Innovation.* Ashland, OR: White Cloud Press, 2013.

Herkewitz, William. "Found: The First Mechanical Gear in a Living Creature." *Popular Mechanics*, September 12, 2013. http://www.popularmechanics.com/science /animals/a9449/the-first-gear-discovered-in-nature-15916433/

Izumi, Hayato, et al. "Realistic Imitation of Mosquito's Proboscis: Electrochemically Etched Sharp and Jagged Needles and Their Cooperative Inserting Motion." *Sensors and Actuators A: Physical* 165:1 (2011): 115–23.

Jallouk, Andrew P., et al. "Nanoparticle Incorporation of Melittin Reduces Sperm and Vaginal Epithelium Cytotoxicity." *PLoS ONE* 9:4 (2014): e95411.

Koerner, Brendan L. "One Doctor's Quest to Save People by Injecting Them with Scorpion Venom." *Wired*, June 24, 2014. https://www.wired.com/2014/06/scorpion -venom/

Latif, Tahmid, and Alper Bozkurt. "Line Following Terrestrial Insect Biobots." *Engineering in Medicine and Biology Society, 2012 Annual Conference of the IEEE* (2012): 972–75.

Lee, Simon, et al. "Cockroaches and Locusts: Physicians' Answer to Infectious Diseases." *International Journal of Antimicrobial Agents* 37.3 (2011): 279–80.

Leu, Chelsea. "Scientists Are Using Tarantula Venom to Learn How Your Body Hurts." *Wired*, June 6, 2016. https://www.wired.com/2016/06/tarantula-toxins-teach-us -science-pain/

Marks, Paul. "Mosquito Needle Helps Take the Sting Out of Injections." *New Scientist*, March 16, 2011. https://www.newscientist.com/article/mg20928044-900-mosquito -needle-helps-take-sting-out-of-injections/

Marquis, Don. *Archy and Mehitabel*. New York: Anchor Books, 1970.

McKenna, Maryn. "The Coming Cost of Superbugs: 10 Million Deaths per Year." *Wired*, December 15, 2014. https://www.wired.com/2014/12/oneill-rpt-amr/

McNeil, Donald G., Jr. "Slithery Medical Symbolism: Worm or Snake? One or Two?" *New York Times*, March 8, 2005. http://www.nytimes.com/2005/03/08/health /slithery-medical-symbolism-worm-or-snake-one-or-two.html

Moore, Malcolm. "Cockroaches: The New Miracle Cure for China's Ailments." *Daily Telegraph*, October 24, 2013. http://www.telegraph.co.uk/news/worldnews/asia /china/10399443/Cockroaches-the-new-miracle-cure-for-Chinas-ailments.html

Morgan, C. Lloyd. "The Beetle in Motion." *Nature* 35:888 (1886): 7.

Nikolic, Vojin. "Low-Sweep and Composite Planform Movable Wing Tip Strakes." 46th AIAA Aerospace Sciences Meeting and Exhibit, January 2008.

Olson, Jim. "Tumor Paint." PopTech 2013, Camden, Maine. https://poptech.org/popcasts /jim_olson_tumor_paint

Piore, Adam. "Rise of the Insect Drones." *Popular Science*, January 2014: 38–43.

"Project Violet: About Us." Fred Hutch. https://www.fredhutch.org/en/labs/clinical /projects/project-violet/about-us.html

Rains, Glen C., Jeffery K. Tomberlin, and Don Kulasiri. "Using Insect Sniffing Devices for Detection." *Trends in Biotechnology* 26:6 (2008): 288–94.

Ratcliffe, Norman, Patricia Azambuja, and Cicero Brasileiro Mello. "Recent Advances in Developing Insect Natural Products as Potential Modern Day Medicines." *Evidence-Based Complementary and Alternative Medicine*, article ID 904958 (2014). https://www.hindawi.com/journals/ecam/2014/904958/

Ray, John. "Concerning Some Uncommon Observations and Experiments Made with an Acid Juice to Be Found in Ants." *Philosophical Transactions* 5 (1670): 2069–77.

"The Rod of Asclepius and Caduceus: Two Ancient Symbols." *Florence Inferno*, June 27, 2016. http://www.florenceinferno.com/rod-of-asclepius-and-caduceus-symbols/

Roy, Spandita, Sumana Saha, and Partha Pal. "Insect Natural Products as Potential Source for Alternative Medicines: A Review." *World Scientific News* 19 (2015): 80–94.

Schmidt, Justin O. *The Sting of the Wild*. Baltimore, MD: Johns Hopkins University Press, 2016.

Schmidt, Justin O., Murray S. Blum, and William L. Overal. "Hemolytic Activities of Stinging Insect Venoms." *Archives of Insect Biochemistry and Physiology* (1984): 155–60.

Scholtz, Gerhard. "Scarab Beetles at the Interface of Wheel Invention in Nature and Culture?" *Contributions to Zoology* 77:3 (2008): 139–48.

Schweid, Richard. *The Cockroach Papers: A Compendium of History and Lore.* New York: Four Walls Eight Windows, 1999.

Sherman, Ronald A., Charles E. Shapiro, and Ronald M. Yang. "Maggot Therapy for Problematic Wounds: Uncommon and Off-Label Applications." *Advances in Skin & Wound Care* 20:11 (2007): 602–10.

Usherwood, James R., and Fritz-Olaf Lehmann. "Phasing of Dragonfly Wings Can Improve Aerodynamic Efficiency by Removing Swirl." *Journal of the Royal Society Interface* 5 (2008): 1303–1307.

Walker, Simon M., et al. "In Vivo Time-Resolved Microtomography Reveals the Mechanics of the Blowfly Flight Motor." *PLoS Biology* 12:3 (2014): e1001823.

Watson, James T., et al. "Control of Obstacle Climbing in the Cockroach, *Blaberus discoidalis*: I. Kinematics." *Journal of Comparative Physiology* 188 (2002): 39–53.

Weiler, Nicholas. "Tarantula Toxins Offer Key Insights into Neuroscience of Pain." *UCSF News*, June 6, 2016. https://www.ucsf.edu/news/2016/06/403166/tarantula-toxins-offer-key-insights-neuroscience-pain

Wood, Robert, Radhika Nagpal, and Gu-Yeon Wei. "Flight of the RoboBees." *Scientific American* 308:3 (2013): 60–65.

Chapter Eight: Executives of Big Bug Biz

The biggest debt of gratitude is owed to Akito Kawahara's fantastic American Entomologist *article subtitled "Entomology in Japan." His research and, later, our informative interview not only aided me in this chapter but inspired a great documentary called* Beetle Queen Conquers Tokyo. *Main sources for this chapter also came from: Debin Ma's paper "Why Japan, Not China, Was the First to Develop in East Asia," Nan-Yao Su's interview and "Tokoyo no kami" paper, and Stuart Fleming's "The Tale of the Cochineal." Here I would also like to bring attention to Hugh Raffle's* Insectopedia. *While it is referenced throughout the book, his coverage on the sport of cricket fighting was a big, awesome part of this chapter.*

Associated Press. "Bolivia, Peru Reject Use of Bugs in Cocaine Fight: War on Drugs: Latin Nations Want to Switch to Legal Crops, Not to Caterpillars or Worms to Eat Coca Leaves." *Los Angeles Times*, February 22, 1990. http://articles.latimes.com/1990-02-22/news/mn-1722_1_coca-leaves

Ballenger, Joe. "Cricket Virus Leads to Illegal Importation of Foreign Species for Pet Food." *Entomology Today*, December 22, 2014. https://entomologytoday.org/2014/12/22/cricket-virus-leads-to-illegal-importation-of-foreign-species-for-pet-food/

Berenbaum, May. "Buzzwords: Just Say 'Notodontid'?" *American Entomologist* 37:4 (1991): 196–97.

Daimon, Takaaki, et al. "The Silkworm *Green b* Locus Encodes a Quercetin 5-O-Glucosyltransferase that Produces Green Cocoons with UV-Shielding Properties." *Proceedings of the National Academy of Sciences* 107:25 (2010): 11471–76.

Eugenides, Jeffrey. *Middlesex.* New York: Picador, 2003.

Evans, Arthur V. *What's Bugging You?: A Fond Look at the Animals We Love to Hate.* Charlottesville: University of Virginia Press, 2008.

Fleming, Stuart. "The Tale of the Cochineal: Insect Farming in the New World." *Archaeology* 36:5 (1983): 68–69, 79.

Gebel, Erika. "Proteins Revealed in Fire Ant Venom." *Chemical & Engineering News,* August 20, 2012. http://cen.acs.org/articles/90/web/2012/08/Proteins-Revealed-Fire -Ant-Venom.html

Govindan, R., T. K. Narayanaswamy, and M. C. Devaiah. *Pebrine Disease of Silkworm.* Bangalore, India: UAS Offset Press, 1997.

Hicks, Edward. *Shellac: Its Origin and Applications.* New York. Chemical Publishing Co., 1961.

Hudak, Stephen. "Jumpin' Jiminy! Virus Silences Cricket Farm." *Orlando Sentinel,* June 21, 2010. http://articles.orlandosentinel.com/2010-06-21/news/os-cricket-farm -bankruptcy-20100621_1_lucky-lure-cricket-farm-virus-bug

Iizuka, Tetsuya, et al. "Colored Fluorescent Silk Made by Transgenic Silkworms." *Advanced Functional Materials* 23 (2013): 5232–39.

Kampmeier, Gail E., and Michael E. Irwin. "Commercialization of the Insects and Their Products." *Encyclopedia of Insects.* Vincent H. Resh and Ring T. Cardé, eds. 2nd edition. Boston: Academic Press, 2009. 220–27.

Kawahara, Akito Y. "Thirty-Foot Telescopic Nets, Bug Collecting Videogames, and Beetle Pets: Entomology in Modern Japan." *American Entomologist* 53:3 (2007): 160–72.

Kolar, Jana, et al. "Historical Iron Gall Ink Containing Documents: Properties Affecting Their Condition." *Analytica Chimica Acta* 555 (2006): 167–74.

"Learning the History." *World Heritage Site Tomioka Silk Mill.* Tomioka City, Japan. http://www.tomioka-silk.jp.e.wv.hp.transer.com/tomioka-silk-mill/guide/history .html

Ma, Debin. "Why Japan, Not China, Was the First to Develop in East Asia: Lessons from Sericulture, 1850–1937." *Economic Development and Cultural Change* 52:2 (2004): 369–94.

McConnaughey, Janet. "Virus Kills Hordes of Cricket Raised for Reptiles." Associated Press, January 12, 2011. http://www.huffingtonpost.com/huff-wires/20110112/us -food-and-farm-cricket-crisis/

Parker, Rosemary. "Following Cricket Paralysis Virus Catastrophe, Top Hat Cricket Farm in Portage Rebuilds Its Business." *Michigan Live,* January 19, 2012. http:// www.mlive.com/news/kalamazoo/index.ssf/2012/01/top_hat_cricket_farm_in _portag.html

Parry, Ernest J. *Shellac: Its Production, Manufacture, Chemistry, Analysis, Commerce and Uses.* London: Sir Isaac Pitman & Sons, 1935.

"Plan to Eradicate Coca Would Use Caterpillars." *New York Times,* February 20, 1990. http://www.nytimes.com/1990/02/20/us/plan-to-eradicate-coca-would-use -caterpillars.html

Raffles, Hugh. *Insectopedia.* New York: Vintage Books, 2010.

Shiva, M. P. *Inventory of Forest Resources for Sustainable Management and Biodiversity Conservation.* New Delhi, India: Indus Publishing, 1998.

"Silkworm Diseases." *The Whole Story.* Institut Pasteur, Paris, France, February 13, 2014.

Su, Nan-Yao. "Tokoyo no Kami: A Caterpillar Worshiped by a Cargo Cult of Ancient Japan." *American Entomologist* 60:3 (2014): 182–88.

Tabunoki, Hiroko, et al. "A Carotenoid-Binding Protein (CBP) Plays a Crucial Role in Cocoon Pigmentation of Silkworm (*Bombyx mori*) Larvae." *FEBS Letters* 567 (2004): 175–78.

Takeda, Satoshi. "*Bombyx mori*." *Encyclopedia of Insects*. Vincent H. Resh and Ring T. Cardé, eds. 2nd edition. Boston: Academic Press, 2009. 117–19.

Theobald, Mary Miley. "Putting the Red in Redcoats." *Colonial Williamsburg Journal*, Summer 2012. https://www.history.org/foundation/journal/summer12/dye.cfm

Tomioka Silk Mill and Related Sites. United Nations Educational, Scientific and Cultural Organization, 2015. http://whc.unesco.org/en/list/1449

Tsurumi, E. Patricia. *Factory Girls: Women in the Thread Mills of Meiji Japan*. Princeton, NJ: Princeton University Press, 1990.

Chapter Nine: Dining with Crickets

In my initial research, one book in particular was extremely helpful: Daniella Martin's Edible. Not only was it a big source triggering a large amount of research, but it served as a recipe book for "wax worm tacos." Additionally, David George Gordon's Eat-A-Bug Cookbook and over-the-phone interview were very helpful, as were Marianne Shockley's interview and her coauthored chapter "Insects for Human Consumption." Without a doubt, another gigantic resource was the United Nations' report titled "Edible Insects," which covers all the benefits of entomophagy.

Ayieko, Monica, V. Oriaro, and I. A. Nyambuga. "Processed Products of Termites and Lake Flies: Improving Entomophagy for Food Security within the Lake Victoria Region." *African Journal of Food, Agriculture, Nutrition and Development* 10:2 (2010): 2085–98.

Chang, David. "The Unified Theory of Deliciousness." *Wired*, August 2016: 78–83.

Demick, Barbara. "Cockroach Farms Multiplying in China." *Los Angeles Times*, October 15, 2013. http://www.latimes.com/world/la-fg-c1-china-cockroach-20131015-dto-htmlstory.html

Gahukar, R. T. "Entomophagy and Human Food Security." *International Journal of Tropical Insect Science* 31:3 (2011): 129–44.

Goodyear, Dana. "Grub: Eating Bugs to Save the Planet." *New Yorker*, August 15, 2011. http://www.newyorker.com/magazine/2011/08/15/grub

Gordon, David George. *The Eat-A-Bug Cookbook*. Berkeley, CA: Ten Speed Press, 1998.

Holt, Vincent M. *Why Not Eat Insects?* London: Field & Tuer, The Leadenhall Press, 1885.

Hongo, Jun. "Waiter . . . There's a Bug in My Soup." *Japan Times*, December 14, 2013. http://www.japantimes.co.jp/life/2013/12/14/lifestyle/waiter-theres-a-bug-in-my-soup/#.WIensbYrKi4

Katayama, Naomi, et al. "Entomophagy: A Key to Space Agriculture." *Advances in Space Research* 41 (2008): 701–5.

——. "Entomophagy as Part of a Space Diet for Habitation on Mars." *Journal of Space Technology and Science* 21:2 (2005): 227–38.

Lee, Nicole. "Grow Your Own Edible Mealworms in a Desktop Hive." *Engadget*, November 11, 2015. https://www.engadget.com/2015/11/11/livin-farms-hive/

MacNeal, David. "Bug Bento." *Wired*, September 17, 2013. https://www.wired.com/2013/09/bugbento/

Martin, Daniella. "The Benefits of Eating Bugs." *The Week*, March 1, 2014. http://theweek
.com/articles/450029/benefits-eating-bugs

——. *Edible: An Adventure into the World of Eating Insects and the Last Great Hope to Save the Planet.* Boston: Houghton Mifflin Harcourt, 2014.

McNeilly, Hamish. "Hungry? Try a 'Sky Prawn' at Dunedin's Vault 21 Restaurant." *Stuff*, May 24, 2016. http://www.stuff.co.nz/oddstuff/80331526/Hungry-Try-a-sky-prawn -at-Dunedins-Vault-21-restaurant

Megido, Rudy Caparros, et al. "Edible Insects Acceptance by Belgian Consumers: Promising Attitude for Entomophagy Development." *Journal of Sensory Studies* 29 (2014): 14–20.

Ramos-Elorduy, Julieta. "Anthropo-Entomophagy: Cultures, Evolution and Sustainability." *Entomological Research* 39 (2009): 271–88.

Raubenheimer, David, and Jessica M. Rothman. "Nutritional Ecology of Entomophagy in Humans and Other Primates." *Annual Review of Entomology* 58 (2013): 141–60.

Shockley, Marianne, and Aaron T. Dossey. "Insects for Human Consumption." *Mass Production of Beneficial Organisms: Invertebrates and Entomopathogens.* Juan A. Morales-Ramos, M. Guadalupe Rojas, and David I. Shapiro-Ilan, eds. Oxford: Elsevier, 2014. 617–52.

van Huis, Arnold, et al. "Edible Insects: Future Prospects for Food and Feed Security." Food and Agriculture Organization of the United Nations. FAO Forestry Paper 171 (2013).

van Huis, Arnold, Henk van Gurp, and Marcel Dicke. *The Insect Cookbook: Food for a Sustainable Planet.* New York: Columbia University Press, 2014.

Wøldike, Christian Korf, Jakub Droppa, and Mads Gustav Grene. "Insects for Dinner: A Study of Entomophagy." Roskilde University. PhD dissertation, 2016.

Yen, Alan L. "Entomophagy and Insect Conservation: Some Thoughts for Digestion." *Journal of Insect Conservation* 13 (2009): 667–70.

Chapter Ten: Tracing the Collapse

Three books were especially informative to this chapter, and those were: Hannah Nordhaus's The Beekeeper's Lament, *Hattie Ellis's* Sweetness & Light, *and coming through with an amazing wealth of historical nuggets Hilda Ransome's* The Sacred Bee. *Thomas Seeley's book* Honeybee Democracy *is largely referenced as well. Several papers coauthored by Dennis vanEngelsdorp, as well as our interview, were very useful, as was my informative interview with CSIRO scientist Paulo de Souza and notes from University of California–Davis professor Elina L. Niño.*

"The Ancient Art of Honey Hunting in Nepal in Pictures." *The Guardian*, February 27, 2014. https://www.theguardian.com/travel/gallery/2014/feb/27/honey-hunters-nepal -in-pictures

Associated Press. "Collateral Damage: Bees Die in South Carolina Zika Spraying." September 1, 2016. http://newsok.com/article/feed/1066424

Cardinal, Sophie, and Bryan N. Danforth. "The Antiquity and Evolutionary History of Social Behavior in Bees." *PLoS ONE* 6:6 (2011): e21086.

Chrysochoos, John. *Ikaria: Paradise in Peril.* Pittsburgh: RoseDog Books, 2010.

Cox-Foster, Diana, and Dennis vanEngelsdorp. "Solving the Mystery of the Vanishing Bees." *Scientific American Magazine*, March 31, 2009. https://www.scientificamerican .com/article/saving-the-honeybee/

Crane, Eva. "Bee Products." *Encyclopedia of Insects*. Vincent H. Resh and Ring T. Cardé, eds. 2nd edition. Boston: Academic Press, 2009. 71–75.

Cresswell, James E. "A Meta-Analysis of Experiments Testing the Effects of a Neonicotinoid Insecticide (Imidacloprid) on Honey Bees." *Ecotoxicology* 20 (2011): 149–57.

Dainat, Benjamin, Dennis vanEngelsdorp, and Peter Neumann. "Colony Collapse Disorder in Europe." *Environmental Microbiology Reports* 4:1 (2012): 123–25.

Eisenstein, Michael. "Seeking Answers Amid a Toxic Debate." *Nature* 521 (2015): 552–55.

Ellis, Hattie. *Sweetness & Light: The Mysterious History of the Honeybee*. New York: Harmony Books, 2004.

Frisch, Karl von. "Decoding the Language of the Bee." Nobel Prize Lecture. University of Munich, Federal Republic of Germany. December 12, 1973. Lecture.

Hornets from Hell. Dir. Jeff Morales. National Geographic Explorer, 2002.

Jansen, Suze A., et al. "Grayanotoxin Poisoning: 'Mad Honey Disease' and Beyond." *Cardiovascular Toxicology* 12:208 (2012): 208–215.

Kantor, Sylvia. "Can Mushrooms Save the Honeybee?" *Crosscut*, February 16, 2015. http://crosscut.com/2015/02/can-mushrooms-save-honeybee/

Laurino, Daniela, et al. "Toxicity of Neonicotinoid Insecticides to Honey Bees: Laboratory Tests." *Bulletin of Insectology* 64:1 (2011): 107–13.

Lenfestey, James P., ed. *If Bees Are Few: A Hive of Bee Poems*. Minneapolis: University of Minnesota Press, 2016.

Maeterlinck, Maurice. *The Life of the Bee*. New York: Mentor Books, 1954.

Main, Douglas M. "A Different Kind of Beekeeping Takes Flight." *New York Times*, February 17, 2012. https://green.blogs.nytimes.com/2012/02/17/a-different-kind-of-beekeeping-takes-flight/

Monk and the Honey Bee, The. Narr. Roger Mills. Dir. Allen Jewhurst and David Taylor. BBC. 1988.

Munz, Tania. *The Dancing Bees: Karl von Frisch and the Discovery of the Honeybee Language*. Chicago: The University of Chicago Press, 2016.

Nordhaus, Hannah. *The Beekeeper's Lament: How One Man and Half a Billion Honey Bees Help Feed America*. New York: Harper Perennial, 2011.

——. "The Honey Trap." *Wired*, September 2016: 70–77.

Pettis, Jeffery S., et al. "Colony Failure Linked to Low Sperm Viability in Honey Bee (*Apis mellifera*) Queens and an Exploration of Potential Causative Factors." *PLoS ONE* 11:2 (2016): e0147220.

Ransome, Hilda M. *The Sacred Bee: In Ancient Times and Folklore*. New York: Houghton Mifflin, 1937.

Seeley, Thomas D. *Honeybee Democracy*. Princeton, NJ: Princeton University Press, 2010.

Seitz, Nicola, et al. "A National Survey of Managed Honey Bee 2014–2015 Annual Colony Losses in the USA." *Journal of Apicultural Research* 54 (2016): 292–304.

"Systemic Pesticides Pose Global Threat to Biodiversity and Ecosystem Services." International Union for Conservation of Nature, June 24, 2014. https://www.iucn.org/content/systemic-pesticides-pose-global-threat-biodiversity-and-ecosystem-services

Tapanila, Leif, and Eric M. Roberts. "The Earliest Evidence of Holometabolan Insect Pupation in Conifer Wood." *PLoS ONE* 7:2 (2012): e31668.

Tison, Léa, et al. "Honey Bees' Behavior Is Impaired by Chronic Exposure to the Neonicotinoid Thiacloprid in the Field." *Environmental Science & Technology* 50 (2016): 7218–27.

Townsend, Gordon F., and Eva Crane. "History of Apiculture." *History of Entomology.* Ray F. Smith, Thomas E. Mittler, and Carroll N. Smith, eds. Palo Alto, CA: Annual Reviews Inc., 1973. 387–406.

Tu, Chau. "Step into a Hive." *Science Friday,* June 29, 2016. http://www.sciencefriday .com/articles/step-into-a-hive/

Weber, Bruce. "Margaret Heldt, Hairdresser Who Built the Beehive, Dies at 98." *New York Times,* June 13, 2016.

Wenner, Adrian M., and William W. Bushing. "Varroa Mite Spread in the United States." *Bee Culture* 124 (1996): 341–43.

Wilford, John Noble. "Which Came First: Bees or Flowers? Find Points to Bees." *New York Times,* May 23, 1995. http://www.nytimes.com/1995/05/23/science/which-came-first -bees-or-flowers-find-points-to-bees.html?pagewanted=all

Woodcock, Ben A., et al. "Replication, Effect Sizes and Identifying the Biological Impacts of Pesticides on Bees under Field Conditions." *Journal of Applied Ecology* 53 (2016): 1358–62.

Index

7/17